DESKBOOK OF BUILDING CONSTRUCTION

Charts, Tables, and Forms

Byron W. Maguire

Prentice-Hall, Inc. Englewood Cliffs, NJ 07632

Library of Congress Cataloging in Publication Data

Maguire, Byron W.,
Deskbook of building construction.

Includes index.
1. Building-Handbooks, manuals, etc. I. Title.
TH151.M26 1985 690'.068 83-24684
ISBN 0-13-202037-8

Editorial/production supervision and
interior design: *Shari Ingerman*
Cover design: *Edsal Enterprises*
Manufacturing buyer: *Tony Caruso*
Page layout: *Jill S. Packer*

Printed in the United States of America

10 9 8 7 6 5 4 3 2 1

ISBN 0-13-202037-8 01

Prentice-Hall International, Inc., *London*
Prentice-Hall of Australia Pty. Limited, *Sydney*
Editora Prentice-Hall do Brasil, Ltda., *Rio de Janeiro*
Prentice-Hall Canada Inc., *Toronto*
Prentice-Hall of India Private Limited, *New Delhi*
Prentice-Hall of Japan, Inc., *Tokyo*
Prentice-Hall of Southeast Asia Pte. Ltd., *Singapore*
Whitehall Books Limited, *Wellington, New Zealand*

CONTENTS

PREFACE

Every construction contractor is in business for several reasons. There is no special hierarchy of reasons, since these are personal to each contractor. Some of the reasons are to make a profit, thereby to obtain wealth or the things money can buy; to become successful, respected, and meet esteem needs; to follow in one's father's footsteps and be as good as he was (My dad said that if I kept at it, I'd be good when I was 50 years old. I believed I was good at 20. He said to keep trying.); to build beautiful objects (houses, bridges, high-rise buildings, gardens, parks, etc.) for others to enjoy for years and years; to provide a legacy for one's heirs—something of worth and substance; and to provide a good living for a team of faithful employees.

To accomplish any one or several of the many reasons for being a contractor, certain well-defined principles or disciplines need to be understood and used. This book covers five of those disciplines: the management of personnel, finances, economics, accounting, and legal responsibilities, as well as several related subjects.

Each of these five disciplines is so broad that colleges and universities all over the world offer degree programs through the doctorate level. Even though as many as four, five, six, or more years are required to obtain degrees, many contractors cannot afford the time or money that is required to complete such programs. Unfortunately, the demands of society and governments are ever-increasing. Thus, to succeed, contractors are required to be more knowledgeable and sophisticated. However, to function effectively and to stay in business, it is vital that every contractor, manager of a contracting firm, superintendent, foreman, and crew chief have knowledge of these disciplines.

This book is written to highlight those parts of each discipline that are of particular importance to the success of all contracting companies. Within each chapter, many, many subjects are addressed, each a short synopsis of a very large segment of the special disciplines. This coverage is broad enough to serve as a reminder for those contractors who occasionally need a refresher of fact or principle. However, the coverage may lack sufficient detail or examples for other contractors. If so, the references which are liberally offered throughout the book provide suggestions for self-study and development, formal education, and group study (seminars). In addition, many associations' and institutions' addresses are listed throughout the book and in an appendix. These should be helpful, as each provides considerable information at very nominal cost.

I have tried to provide a very broad but thorough coverage of every aspect of construction contracting that relates to profit making, company integrity, and the interaction between the contractor and his or her employees and other people in the business sphere. It is my sincere wish that this book will prove valuable to you for many years and will, in some small measure, add to your success and personal growth.

B.M.

1

GENERAL MANAGEMENT CONCEPTS

QUICK REFERENCES

This subject of general management concepts provides a variety of data that relate to fundamentals that all contractors and supervisory personnel need to know if they are to be effective. Together with *theories of management*, you will find data on scientific, administrative, and behavioral management concepts. The *elements of management* include planning, organizing, directing, controlling, coordinating, and *organizational structures*. Finally, several *work-flow structures* are demonstrated.

Most of this information will serve as a refresher to managers who have had considerable formal training. Rereading it may initiate recall and bring forth ideas that will have immediate application. Other readers may find in this information new insights leading to reflective thinking on a current problem, one that hitherto has had no easy or apparent solution. In either event, readers can refer to the section that supplies immediate material to help resolve a known or suspected need.

THEORIES OF MANAGEMENT

Scientific Management[1]

Scientific management as defined by Frederick Taylor centered around the idea that blue-collar work could be studied scientifically. He spent considerable time and effort using his engineering background to formulate his theories. Taylor proposed a new approach to formulate his theories for managers to look at the workplace, and in *Shop Management and The Principles of Scientific Management*, he suggested four principles:

1. Find the one best way to perform each job by careful observation.
2. Scientifically select personnel and develop them to their most efficient capacity.
3. Provide financial incentives according to a predetermined value per job effort.
4. Create functional foremanship, thereby creating a division between management and workers.

We can easily see the immediate application of Taylor's principles to every construction job. As manager, entrepreneur, foreman, or section boss, we always attempt to define the one best way to perform each job assignment. Further, we always attempt to proceed in a logical building-block sequence. Second, the crew needed to perform the task or series of tasks will need special, well-defined skills. Therefore, personnel selection incorporates the second principle. Third, in the past, wages were set on a piece work basis. For example, each brick laid earned x cents; each door hung also earned x cents. Today, hourly wages are paid, and incentive pay for early completion is frequently offered. Fringe benefits are also used as incentives.

Creating functional foremanship is also used. Consider the entrepreneur; he or she may act as a general contractor and foreman, planning, directing, and coordinating activities for the workers. Or, he or she may reserve the managerial responsibilities and allow a foreman to guide and direct the workers throughout each job. The workers follow these orders and perform a variety of tasks.

[1]Adapted from Gary Dessler's discussion [*Organization Theory: Integrating Structure and Behavior* (Englewood Cliffs, N.J.: Prentice-Hall, 1980), p. 17] of Frederick Winslow Taylor's basic *scientific management* concepts developed in the early 1900s. Used by permission of Prentice-Hall, Inc., Englewood Cliffs, N.J.

Administrative Management

Henry Fayol, a Frenchman, produced a work entitled *General and Industrial Management* in which he stated that all industrial activities could be placed into one of six groups[2]:

1. Technical activities (production, manufacture, adaption)
2. Commercial activities (buying, selling, exchange)
3. Financial activities (search for the optimum use of capital)
4. Security activities (protection of property)
5. Accounting activities (balance sheets, profit-and-loss statements, statistics)
6. Managerial activities (planning, organizing, command, coordination, and control)

There is no doubt that each of these groups is a function of administrative management. The question of applicability depends on the position one holds in a firm. The proprietor needs involvement in all six groups if he or she expects to make a profit year after year.

On the other hand, a contracting firm that is corporate in structure may, and probably will, employ specialists in all six areas. Or possibly, the owners—the stockholders—will manage the firm.

Today, managers, whether they are proprietors, partners, or general managers, employ Fayol's 14 principles of management:

1. *Division of work*: The contracting manager manages, and the workers perform their skills by repeating their tasks; both groups get better and better.
2. *Authority and responsibility*: The managers and foremen give orders and have the power to extract obedience—this is authority. Responsibility is accepting the consequences for actions taken by managers.
3. *Discipline*: The giving and withholding of rewards for service or work, such as bonuses, promotions, and material rewards.
4. *Unity of command*: This is better known as the chain of command—or one boss for each worker.
5. *Unity of direction*: All personnel should be working toward accomplishing the firm's objective.
6. *Subordination of individual interests to general interests*: This means that employees work for the interests of the firm, subordinating their personal interests to those of the firm. However, where a craftsman's interest is applying his talents to the degree that he believes is required, the firm that allows him to do so gains without active subordination.

[2]Ibid., pp. 19–21. The author has adapted Henry Fayol's 14 points to apply to the construction industry. Henry Fayol was a French executive who spent considerable time developing *administrative management* theories.

7. *Remuneration of personnel*: These are pay and allowances, incentives, and fringe benefits paid to employees consistent with the employee's efforts and the firm's ability to pay.
8. *Centralization*: This means that decision making about the organization of work for the managers, any divisions, and all the workers is usually controlled by the highest level in the organization.
9. *Scalar chain*: The ranking of workers and managers from bottom to top.
10. *Order*: The worker's place and the manager's place in the firm.
11. *Equity*: Treatment of the firm's personnel such that they are respected and in turn display loyalty and sincerity.
12. *Stability of tenure of personnel*: The creation of an environment that allows time for personnel to mold to the job, then promoting from within to the degree possible within the capability of the employee (e.g., rough carpenter to finish carpenter to foreman).
13. *Initiative*: The use by employees of the firm's capabilities to plan and carry out plans consistent with the objectives of the firm.
14. *Esprit de corps*: The employees' attitude toward the firm, involving harmony, inner strength, and trust.

Fayol's 14 points are extremely pertinent to the contractors and the management of their firms. Each point, if employed in managing, should produce a particularly favorable work environment. It should be one that is easy to control administratively.

Behavioral Management

Theories X and Y. Douglas McGregor's theories X and Y of interaction within a firm suggest that persons react according to tradition and nature. *Theory X* holds that most people must be closely supervised, controlled, and coerced into achieving a firm's objectives.[3] There is little doubt that much of this attitude prevails within the construction industry: for example, employees taking unauthorized shortcuts when not supervised closely; others deliberately slowing down the work, causing excessive waste or performing sloppy, substandard work; or the paying of extremely high wages to employees because of the work environment.

However, McGregor also formulated *Theory Y*,[4] which assumes that people enjoy work, and if conditions are favorable, will exercise self-control over their performance. This theory postulates that people are motivated to do a good job and work well with peers, subordinates, and supervisors rather than simply working for wages: for example, workers finish pouring and smoothing concrete regardless of the time; or foremen, specialists, and skilled

[3]Douglas McGregor, *The Human Side of Enterprise* (New York: McGraw-Hill, 1960), p. 33, adapted from his Theory X: "The average human being has an inherent dislike of work and will avoid it if he can."

[4]Ibid., pp. 46, 47; adapted from McGregor's assumptions of Theory Y.

journeymen take plans for the next phase of construction or day's work home to study, without concern for pay.

Some of McGregor's ideas are worthy of serious consideration by the construction industry. He advocates several ideas that apply to our needs as managers.

1. *Decentralization and delegation.* Here we can allow a great deal of local freedom with minimum control, so that workers can perform that which they are best suited to do.
2. *Job enlargement.* Here we can allow employees to gain experience in a variety of subskills and skills. For example, rotating workers from rough-framing crews to outside-finishing crews increases their self-worth and the company's worth.
3. *Participative management.* Here workers are asked to provide inputs that will increase well-being on the job site, improve the work flow, reduce safety hazards, and others.
4. *Management by objectives.* Here the workers meet, and understanding the goals of the firm, establish milestones (MBOs) which they feel that they can achieve through their concerted effort.

McGregor sheds some light on management practices and allows us a way of improving the job site as well as interaction among employees and management.

Motivator-Hygiene Theory. A second management theory that concentrates on the human or personal aspects is Frederick Herzberg's motivator-hygiene theory. He identifies "hygiene factors" as those which, if absent, make employees feel "exceptionally bad." These are extrinsic in that they come from outside the person, but they prevent dissatisfaction when they are adequate. Herzberg called the factors that led to satisfaction "motivator factors." If present, these make the employees feel "exceptionally good." These come from inside the person and lead to motivation when managers build them into their operations.[5]

Figure 1-1 illustrates Herzberg's theory. We can easily adapt these ideas into manager–employee relationships in the construction arena. For example, job security is always a problem in that jobs are for the most part short-lived. However, managers frequently keep, intact, crews of craftsmen and staffs.

In another area, supervision, management provides a pecking order; but foremen and crew chiefs seldom tell workers *how to do the work.* They almost always tell them the job to be done. The worker knows his trade or job and proceeds accordingly. On the other hand, if no supervision existed, the employees would have interchanges among themselves and with resulting dissatisfaction.

[5]Frederick Herzberg, *The Motivation to Work* (New York: Wiley, 1959).

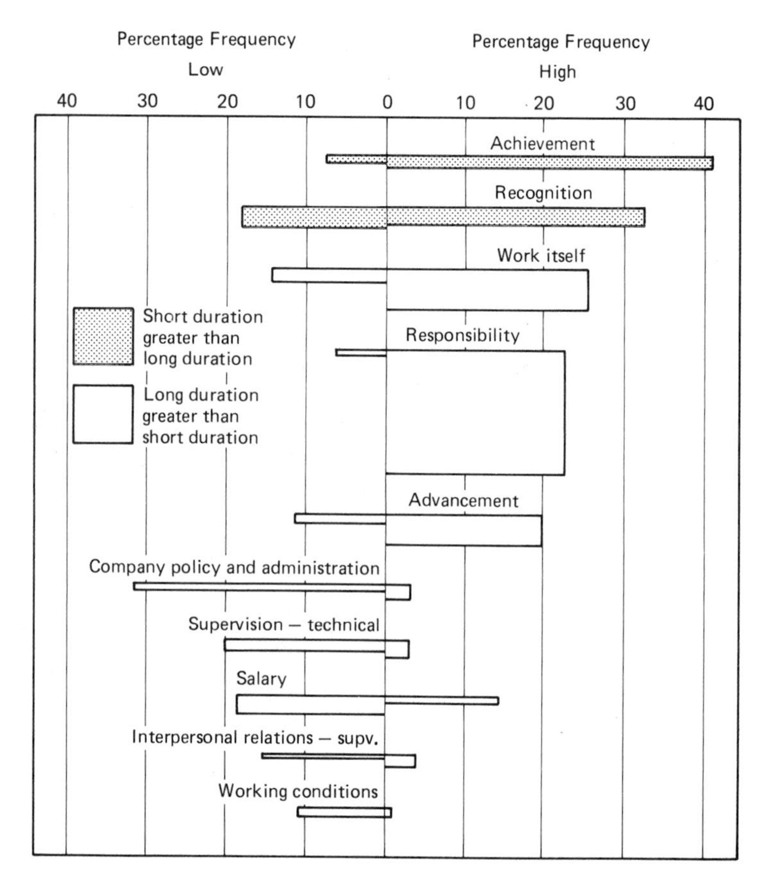

Figure 1-1 Comparison of satisfiers and dissatisfiers in Herzberg's motivator-hygiene theory of management. [Reproduced with permission from F. Herzberg, *The Motivation to Work* (New York: John Wiley & Sons, Inc., 1959).]

The model shows that the best motivating factors are growth, advancement, responsibility, work recognition, and achievement. Managers who place into being situations, or create conditions, where their workers can work toward these factors usually have smooth operations and make money.

But as with all theories, there may be flaws and times where application seems impossible. Further study should help the reader consider some aspects of this motivational management theory because it is adaptable to the personnel situations confronting managers in the construction industry.

Organic and Mechanistic System of Management

Burns and Stalker conducted studies concerning the environment and the organization. They concluded that there are two divergent systems of management practice, the organic and the mechanistic.[6] What are these, and how do they fit the construction site and management?

[6]Burns and Stalker, *The Management of Innovation* (New York: Methuen, Inc., 1961), pp. 120-122.

Before we look at some of the characteristics of each, let us define each type by example. Where the manager adopts the *organic* approach, he or she recognizes that knowledge about the construction job and all of its ramifications resides among the workers. They possess the variety of knowledge whose sum is sufficient to solve all decisions necessary to completing the project. The manager does not possess total knowledge of the job, in that he or she cannot perform the hundreds of individual tasks. This then leads to horizontal communication, consisting of information and advice from workers specialized in certain areas to other workers. Commands are seldom, if ever, used. Commitment by the workers increases significantly toward the concerns, objectives, and goals of the firm. In application, masons, carpenters, steelworkers, plumbers, electricians, earth-moving personnel, and surveyors communicate back and forth regarding the needs of the project, each contributing his special knowledge.

In contrast, a manager adopting the *mechanistic* system of management would stress specialized differentiation of function wherein the total job is broken down into tasks. Workers are hired to complete assigned tasks with little or no concern for tasks out of their immediate realm. Here, precise rights, obligations, and technical methods are attached to each functional role the worker is assigned. There is a strict hierarchy of control; communication is vertical and directive. Worker infringement on management and other workers' responsible functions is looked upon as a threat and frequently precipitates strikes.

Could a firm employ either system depending on the nature of the contract project? The answer, of course, is yes. If a project is to build 25 condominiums, management may adopt the mechanistic system by establishing specialized crews who will complete their individualized tasks in the design stream or work flow. They may be:

1. The foundation crew
2. The framing crew
3. The rough-plumbing crew
4. The rough-electrical crew
5. Exterior-finishing crews
6. Roofing crews
7. Heating-installation crews
8. Painting crews
9. Landscape crews

Within this organization, organic systems by managers can be employed. The superintendent and all crew chiefs can be allowed to operate free of the firm's directives except those key elements reserved for management: schedules, hiring policies, and wage scales.

Management by Objectives (MBOs)

Management by objectives is not a new concept. It has been around since the second quarter of this century and has been in practice much longer than that. It does have applicability to the construction industry.

Management by objectives means, simply, that an objective or series of objectives are defined and written down. Then downstream target completion dates are set. Finally, the workers and managers strive to complete the objectives as scheduled.

The difficulty in using MBOs is that the workers must have had direct involvement in their formation and they must also perceive the objectives as part of their own satisfiers and goals. Anything short of this means that an MBO is nothing more than a company directive. Two items become clear. MBOs are usually well developed and accepted by workers if they are short term. For example, MBOs on a single residential home project could be set by workers because of a company incentive program. They are easily traceable and involve all workers. The workers benefit and management benefits. The second is that MBO must be timely and stimulate the workers toward their personal goals. If, for example, an MBO was to have workers participate in company profits at year's end, based on the workers' efforts, the MBO would fail, because it is likely that many workers would not be with the firm at year's end. It is also possible that a worker's goal may be self-development and promotional opportunity, not a share in profit that may or may not materialize.

In essence, then, MBOs need to:

1. Be developed with worker inputs and acceptance
2. Be short term versus long term
3. Meet the need or needs of each worker who is expected to participate
4. Benefit the company in a specific way that promotes its objective or goal

Now let us extend our thoughts from the job site to the more complex organization. MBOs are useful here as well. However, there is no lessening of commitment by personnel. Division managers may meet and establish MBOs for their divisions. Department heads can do likewise, and the board of directors or partners can set company goals as MBOs. In these situations the MBOs may well cover much longer time spans than those mentioned earlier. Sometimes it takes years to gain a certain percent of the market in specialization. Other long-term MBOs may involve increasing efficiency and stability. Another could be to reduce long-term debt or increase retained earnings. In these situations the people involved expect to have long-term commitments with the firm and expect to satisfy some of their personal goals by commiting themselves to the projects. Mechanically, MBOs are drawn up in chart form (see Exhibit 1-1) and posted where the personnel involved can see the progress being made.

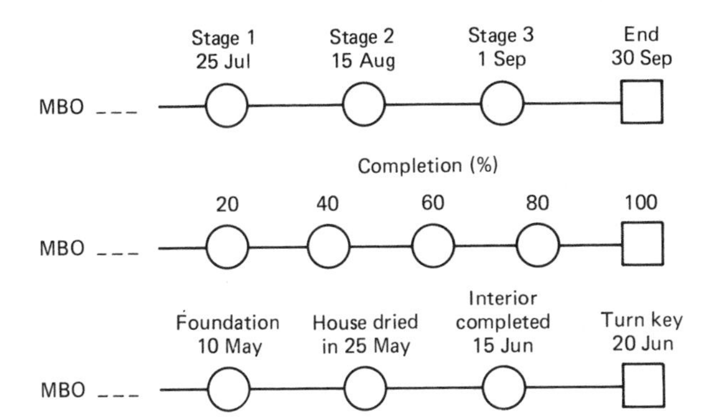

Exhibit 1-1 MBOs: management by objectives.

These short exposures to general management have only touched on scientific, administrative, and behavioral motivational aspects and MBOs. Needless to say, there are many, many more aspects which we have not discussed. Each presents us with a new twist in getting along productively on the job site. Owners, managers, and leaders are more effective in our changing society by continuing to expand their knowledge, and to blend the new information with proven results from the past to achieve desired ends.

REFERENCES AND SOURCES

1. Styles of leadership—considerate, structuring, production oriented, close supervision, and sensitive-awareness types.
   a. Ohio State University studies, Bureau of Business Research. (various studies)
   b. University of Michigan studies, The Survey Research Center. (various studies)
   c. Fiedler, Fred E., *Theory of Leadership Effectiveness* (New York: McGraw-Hill, 1967).
   d. Participative leadership; Katz, Daniel, and Kahn, Robert L., *The Social Psychology of Organizations*, 2nd ed. (New York: Wiley, 1978).
   e. Likert, Rensis, *New Patterns of Management* (New York: McGraw-Hill, 1961).
   f. *Academy of Management Journal*, c/o Dennis Ray, Box KZ, Mississippi State, MS 39762
   g. *Academy of Management Review*, c/o Dennis Ray, Box KZ, Mississippi State, MS 39762
   h. American Management Association, 135 W. 50th St., New York, NY 10020
2. Group influences on management's effectiveness.
   a. George C. Homan's model in his *The Human Group* (New York: Harcourt Brace Jovanovich, 1950).

b. Hare, A. Paul, *Handbook of Small Group Research*, 2nd ed. (New York: Free Press, 1976).

3. Conflict, types, sources, models, managing conflict.
   a. Dessler, Gary, *Organization Theory: Integrating Structure and Behavior* (Englewood Cliffs, N.J.: Prentice-Hall, 1980).
   b. Robbins, Stephen P., *Managing Organizational Conflict: A Non-Traditional Approach* (Englewood Cliffs, N.J.: Prentice-Hall, 1974).
   c. Litterer, Joseph A., "Conflict in Organization," *Academy Management Journal.*
   d. Dalton, M., "Changing Line–Staff Relations," *Personnel Administrations*, April 1966.
   e. Pondy, Louis R., "Organizational Conflict."
4. MBOs
   a. Drucker, Peter, *Practice in Management* (1954).
   b. McGregor, Douglas, *The Human Side of Enterprise* (New York: McGraw-Hill Book Co. Inc., 1960).
   c. Dale, Ernest, and Michelon, L. C., *Modern Management Methods* (New York: World Publishing, 1966).
   d. Schleh, Edward C., *Management by Results: The Dynamics of Profitable Management* (New York: McGraw-Hill, 1961).

ELEMENTS OF MANAGEMENT

All firms, whether a proprietorship, partnership, or corporation, employ the five elements of management. We discuss these next.

Planning

Planning by enterprise's leaders is similar in some cases, regardless of the type of organization, and different in others, depending on who within the organization makes the planning decisions. We shall define several.

In the *sole proprietorship*, the proprietor makes the planning decisions. Since most of these organizations are informally structured, the plans frequently are to:

1. Set short- and long-range goals.
2. Obtain sufficient contracts to remain in business.
3. Make sufficient income to sustain self and family.
4. Make a profit to increase the worth of the company.
5. Plan for major equipment purchases.
6. Plan for efficient use of time, employees' skills, and materials.
7. Plan to employ all tax advantages consistent with the firm's stature.

In a partnership the planning decisions are made by the partners. Frequently, partners specialize in one or more aspects of the company's operations. Therefore, they each contribute to planning consistent with their special skills. Some planning decisions include:

1. Set goals and objectives for the company.
2. Establish guides for each partner's activity within the company.
3. Plan for the distribution of income and retained earnings, growth, expansion, and so on.
4. Plan the operation, number of employees, types of contracts desired, and divisions of the company.

In a *limited partnership* the planning decisions are usually relegated to the general manager subject to approval by the partners (and stockholders). In small organizations one of the partners may assume the general manager's role. However, this does not have to be the case.

The partners usually establish the initial planning and the general manager does the day-to-day planning. Some considerations are:

1. For the partners:
   a. Set company goals and objectives.
   b. Establish guides for partners' and general manager's functions.
   c. Plan for distribution of income.
   d. Plan for large-scale equipment purchases.
2. For the general manager:
   a. Plan the organizational structure.
   b. Develop the pecking order.
   c. Define the degree of formality.
   d. Propose projects to the partners.
   e. Plan the day-to-day operations.

In a *corporation* the stockholders are the owners, and a board of directors make planning decisions. The stockholders usually relinquish planning decisions to the board. They show their pleasure with the board's decisions by reelecting the board members, and displeasure by voting them out of office. The board's planning considerations include:

1. Setting the firm's goals and objectives
2. Planning for growth of the company
3. Planning for acquisition of capital
4. Planning for use of retained earnings
5. Establishing centralization and decentralization criteria
6. Establishing the divisions of the firm

Planning is an absolute requirement for every form of contracting firm. It must be done and done well. If the firm's leadership senses that a limited capacity is all there is, consulting services should be employed to aid in planning.

Given the economic situation as an influence, the firm's plans will need to be modified. This is always fundamental, and when overlooked, firms frequently fail.

Organizing

Organizing the firm usually falls along traditional lines. Some firms start by being informally organized and gradually advance into formalized structures. Others remain informal during their entire life. Still others begin formally and expand or grow and never give up their informal structure.

The essence of organizing can therefore be simplified into several clear steps:

1. Define the jobs to be performed by employees consistent with the objectives of the company.
2. Establish a hierarchy of order (a pecking order) that provides for levels of authority and responsibility.
3. Obtain workers to fill those positions.

We have already identified several organizational schemes or structures in the section on planning. The sole proprietor traditionally organizes the total operation informally, centered around the proprietor. All workers have direct communication with this person. Support roles played by workers often overlap. As a result, there usually is no clear line of authority below the entrepreneur. However, as the organization grows, one or more foremen are employed with stricter responsibilities and lines of authority. One step further and crew chiefs or gang bosses are hired to specialize in certain construction schemes. Further, staff people, such as estimators, accountants, and office workers who have very specialized authority and responsibility may be employed.

Directing

Another way of expressing directing is *command*. But how shall directing be accomplished? Should we employ McGregor's Theory X as one side of a continuum or Theory Y at the other end and direct the firm somewhere between the two? Or should we use Herzberg's motivator-hygiene factors, and if so, to what extent do we wish to direct through company policy and administration?

There are many theories on directing the efforts of workers. Some of the latest theories advocate providing just enough direction to meet successfully the objectives of the firm. Again, though, the size and design of the organization plays a significant role in directing.

1. *Small numbers of employees:* Informal spoken direction, few or no written directives
2. *Sections of horizontal and vertical structures:* Informal direction within sections; semiformal to formal written directives between sections and subsections
3. *Divisions:* Formal directives to sections and other divisions under the parent division
4. *Managers of the firm:* Formal direction downward; informal between managers and staff

One more point on directing. If workers' jobs are clearly identified and the various tasks are cited in job descriptions, everyone in the firm has a much better understanding of his or her individual position. Job separations, overlaps, and hierarchy are easily understood. This type of groundwork by management makes directing much, much easier. Complaints, grievances, and other problems involving employees and affecting job processes are more easily resolved.

Coordinating

Every firm's ability to achieve its goals or objectives is directly proportional to its degree of coordination. Coordinating requires that all members of a firm and those associated with the firm but not of the firm be consistent with the goals. Coordinating activities differ where firms are structured differently.

The *entrepreneur*, operating informally, coordinates all activities personally. Employees receive their instructions daily; suppliers are constantly called to assure delivery of materials to prevent work stoppages. Subcontractors are notified when they are to begin their phase of the total job.

Naturally, coordinating requires two-way communication. Therefore, workers, suppliers, and subcontractors need to inform the contractor of any deviation from their planned activities within the job.

Partners in a *partnership* coordinate constantly because the success of the firm requires it. This is true even when one partner acts as general manager.

Each partner coordinates the activities under his or her domain much like the entrepreneur. Let us assume that partner A is responsible for general management and that partner B is responsible for production. Partner A has accounting, office, sales, and estimating functions. He or she coordinates all the activities and interactivities of these staff functions.

Partner B has several crews working on different jobs and schedules subcontractor work. He or she coordinates the work flow assignments and handles coordination with suppliers.

Both partners coordinate the firm's direction, objectives, and goals. They decide what jobs they want to bid for, what image they want to project, the geographical area in which they wish to operate, and the degree of specialization or generalization they believe will permit them to remain in business successfully.

Coordination in the *corporation* is very much like putting the entrepreneurship and partnership together under one umbrella. Coordinating must be done vertically between levels in the corporation as well as horizontally from division to division or section to section. Managers that adopt open coordination policies generally have smoother-running, more profitable operations than those who use dictatorial downward routes of communication.

In summary, coordinating the activities of a firm is a serious business, and success or failure may hinge on this part of management.

Controlling

To control there must be two elements: (1) possession of the power and authority, and (2) a plan of operation that has a focal point and is workable.

The proprietor, partners, and stockholders of corporations possess the power and authority of the respective firms. They are the ones who invested in the firms. They may retain the authority, as a proprietor and partners usually do. Or they may vest the authority in a general manager or board of directors, or both, as large companies do.

Controlling involves all levels within the firm. Foremen control the work activities of the workmen; gang bosses do likewise. The firm's employees control the scheduling for subcontractors' work and suppliers' deliveries. They also control the flow of work by work-flow schedules. They control spending by constantly keeping tabs on job expenses as compared to scheduled completion dates. Accountants control spending, borrow from banks when needed, pay back to reduce interest costs, project taxes, and perform related controls over expenses. Office personnel control the flow of correspondence within the firm so that managers have useful, current data on various parts of each contract.

In summary, every contracting firm employs the five elements of management. To do less threatens the success of the firm. To employ them effectively increases the firm's success. Informally structured firms usually do not write down these elements formally and then set them in motion. But large-scale, formally structured firms frequently do put these elements in writing as company policy, administrative policy, or operating guides and rules.

REFERENCES

1. Fayol, H., *General and Industrial Management* (New York: Pitman Publishing Corp., 1967).
2. Simon, Herbert A., *Administrative Behavior*, 3rd ed. (New York: Free Press, 1976).

3. Dessler, Gary, *Organization Theory: Integrating Structure and Behavior* (Englewood Cliffs, N.J.: Prentice-Hall, 1980).
4. Dale, Ernest, and Michelon, L. C., *Modern Management Methods* (New York: World Publishing, 1966).

ORGANIZATIONAL STRUCTURE

All forms of contracting firms in the construction industry fall into one of three basic types: sole proprietorship, partnership, or corporation. Within the three classes, subclasses can occur, and these will be demonstrated in flowcharts later in this section.

Line and Staff

It is important to understand that firms are organized to be efficient in their operations, for control, and to establish lines of communication and authority. This boils down to *line* or *staff* or *line and staff* organizational structure. A *line organization* (Figure 1-2a) is one in which there is either one-way communication (down) or two-way communication (up and down). All the functions (elements) of managing are conducted in a line from top (manager) to bottom (lowest worker, apprentice, or laborer). A *staff organization* is one that answers to a central head (Figure 1-2b). Each staff agency has a set of objectives unique to its operation or purpose. All results produced by the staff are channeled to the head (managers). Purely staff organizations are infrequently used in the construction industry compared with line and line and staff. Naturally, a *line and staff* organization includes both *line* (vertical)

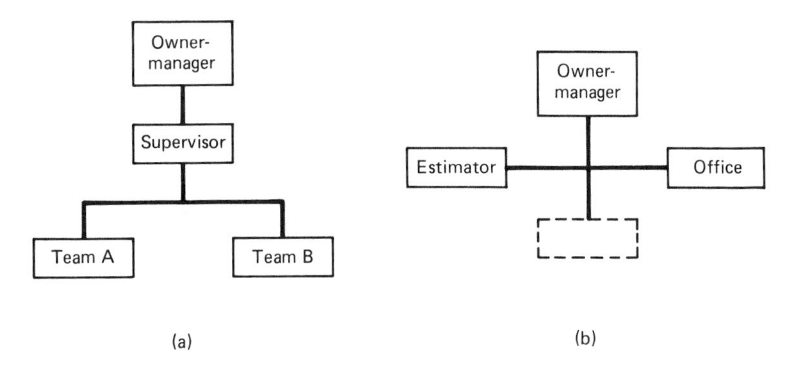

Figure 1-2 Line (a) and staff (b) organization charts.

communications and lines of authority as well as support elements *staff*. Almost all companies in the construction industry employ line and staff organizational structure.

Structural Forms

We shall now examine each type of firm separately, beginning with the sole proprietor.

Sole Proprietorship

1. Informal sole proprietorship with one or more workers (Figure 1-3)

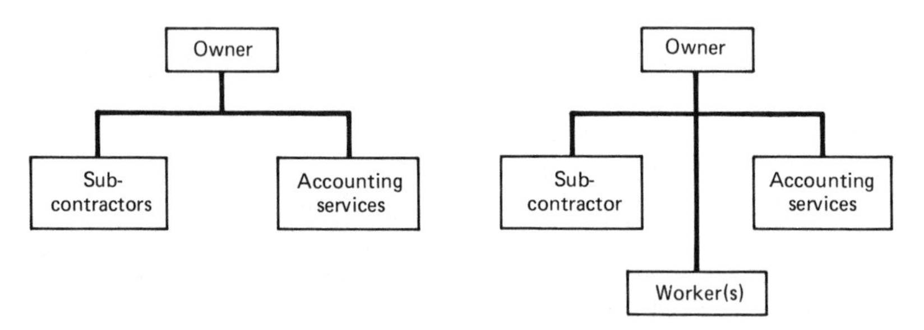

Figure 1-3 Organization chart of an informal sole proprietorship with one or more workers.

2. Informal with paid employees, foremen, and office staff (Figure 1-4)

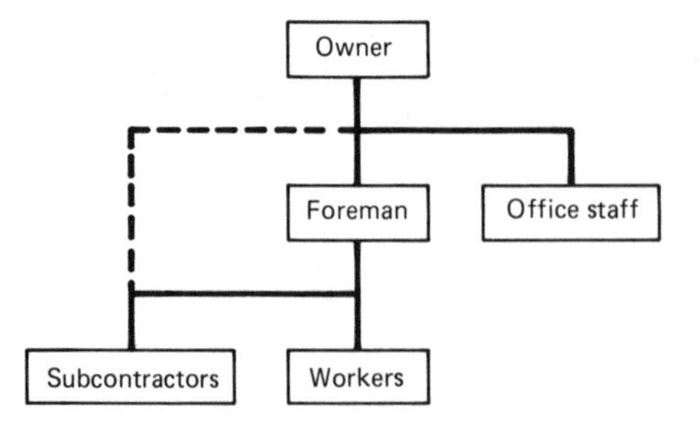

Figure 1-4 Organization chart of an informal sole proprietorship with paid employees, foreman, and office staff.

3. **Formal with multiple jobs and various paid employees (Figure 1-5)**

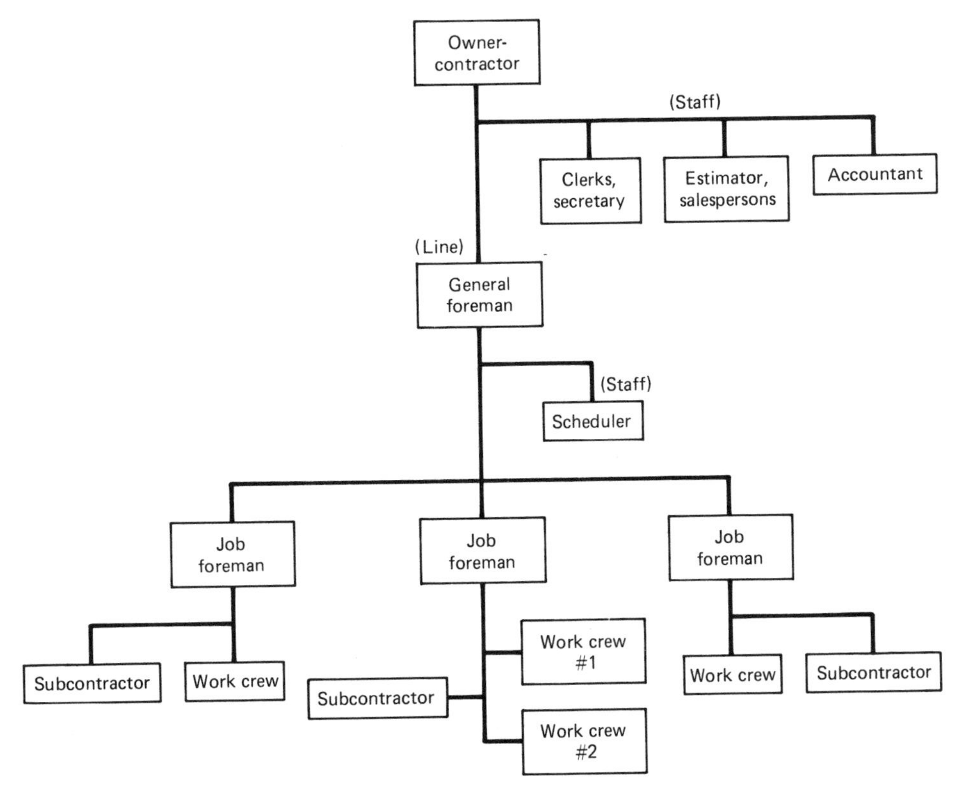

Figure 1-5 Organization chart of a formal sole proprietorship with multiple jobs and various paid employees.

Partnership and Limited Partnership

1. **Informal-formal (Figure 1-6)**

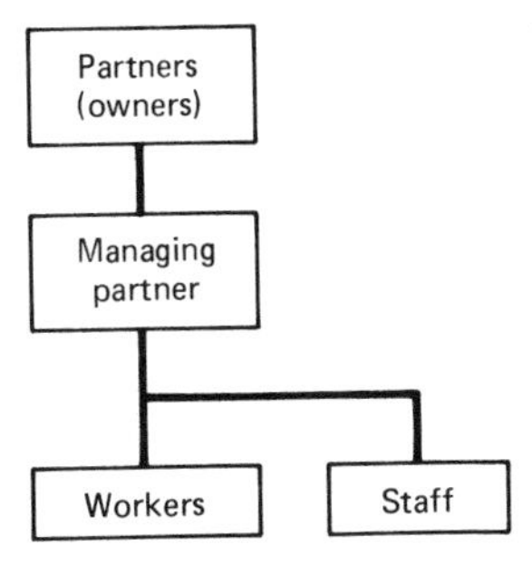

Figure 1-6 Organization chart of an informal-formal partnership. *Note*: Second partner may be foreman.

2. Formal (Figure 1-7)

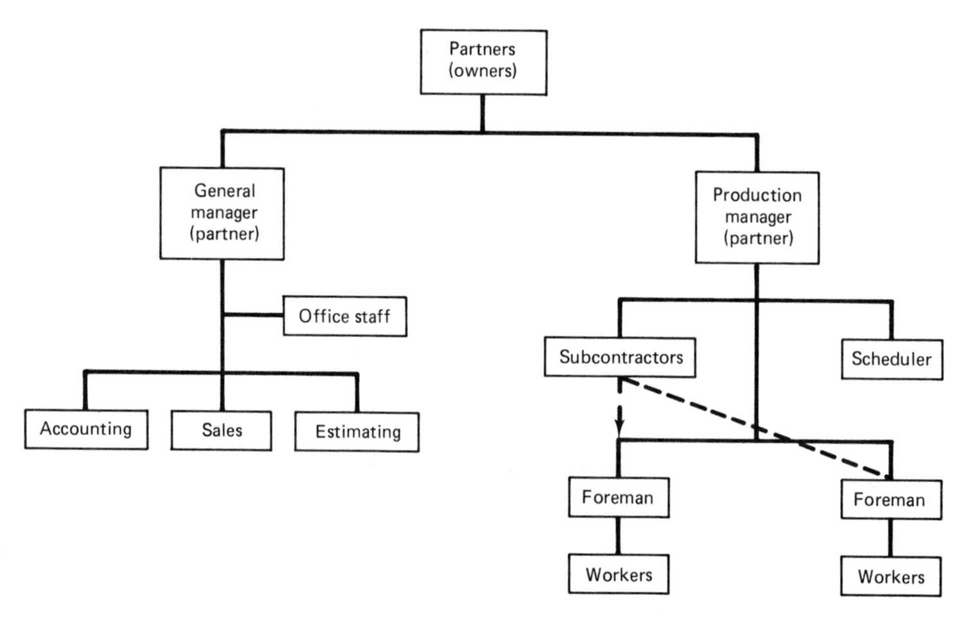

Figure 1-7 Organization chart of a formal partnership.

3. Limited partnership (Figure 1-8)

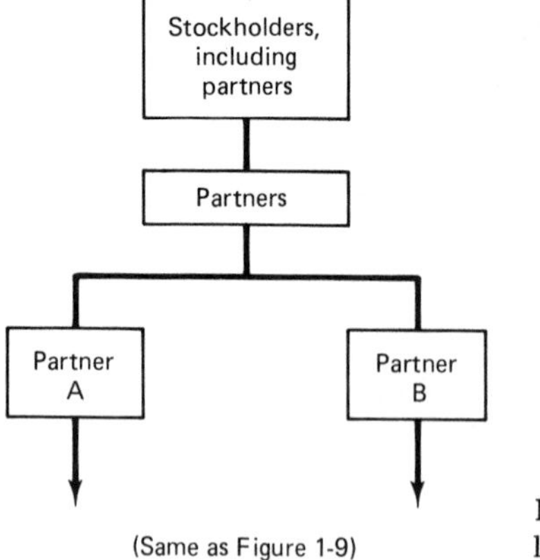

Figure 1-8 Organization chart of a limited partnership.

Corporation

1. Formal (Figure 1-9)

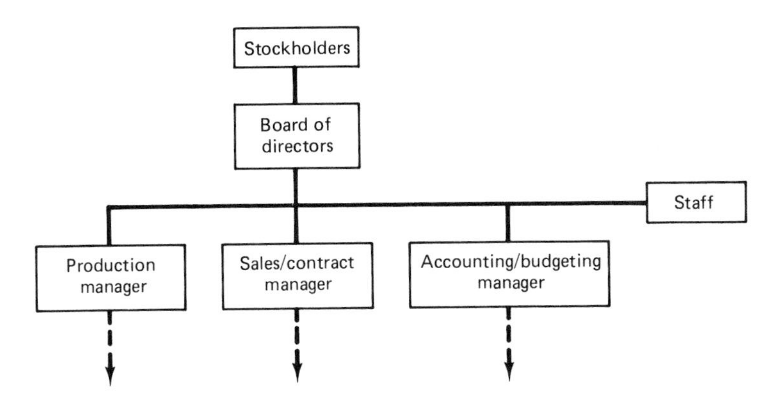

Figure 1-9 Formal corporation.

2. Formal with geographical separation (Figure 1-10)

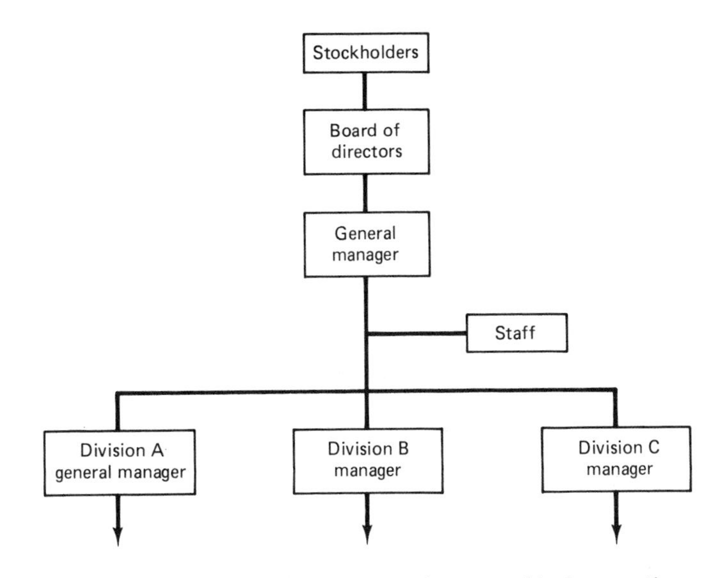

Figure 1-10 Formal corporation with geographical separation.

Each division acts like a separate firm under the restrictions placed on it by the board of directors. The autonomy given to each division is a function of the board. It may opt for centralization, where all decisions are passed down to the division managers, or it may opt for decentralization, where division managers are almost autonomous. Or selected authority is withheld by the board of directors, but considerable freedom is given to each division manager. All of these forms of organization are found in firms engaged in construction.

REFERENCES AND SOURCES

1. Dessler, Gary, *Organization Theory: Integrating Structure and Behavior* (Englewood Cliffs, N.J.: Prentice-Hall, 1980). (Has a good section on both line and staff theory with descriptions of problems and solutions.)
2. Related subjects
   a. New patterns in management
   b. Organization theory (various authors)
   c. Organizational development
   d. Contemporary organizations
   e. Organization's development affected by the environment
3. Dale, Ernest, *Organization* (American Management Association, 1967).
4. Chandler, Alfred, *Strategy and Structure* (Cambridge: M.I.T. Press, 1962).

WORK-FLOW STRUCTURES

Up to now we have been examining the general subject of management. It is certainly a very complex subject, especially when we must use it to make a profit, cause a company to grow, interact with a new market, satisfy employees' needs, and for the myriad of other goals or objectives a company may have. Eventually, though, all managers must come to grips with the day-to-day operations and, of course, this includes the flow of the job.

Single-Job Work Flow

The work-flow structure for a *single job* would be viewed as follows. Suppose that the contractor must build a residential house. (See table on p. 21.)

This *big* picture begins the work-flow structure but by no means provides the decision blocks necessary to assure satisfactory job completion. We shall look at one technique that can be applied.

Figure 1-11 combines phase I and II into several critical blocks. In

Phase	Job	Done by:
I	Ground preparation	Subcontractor
II	Foundation preparation	Carpenter, masons, plumber
III	Erection	Carpenters
IV	Installation of utilities	Subcontractors
V	Finishing	Carpenters, masons, painters
VI	Landscaping	Subcontractor
VII	Inspection–adjustments	Prime and subcontractors, bank, federal inspection service

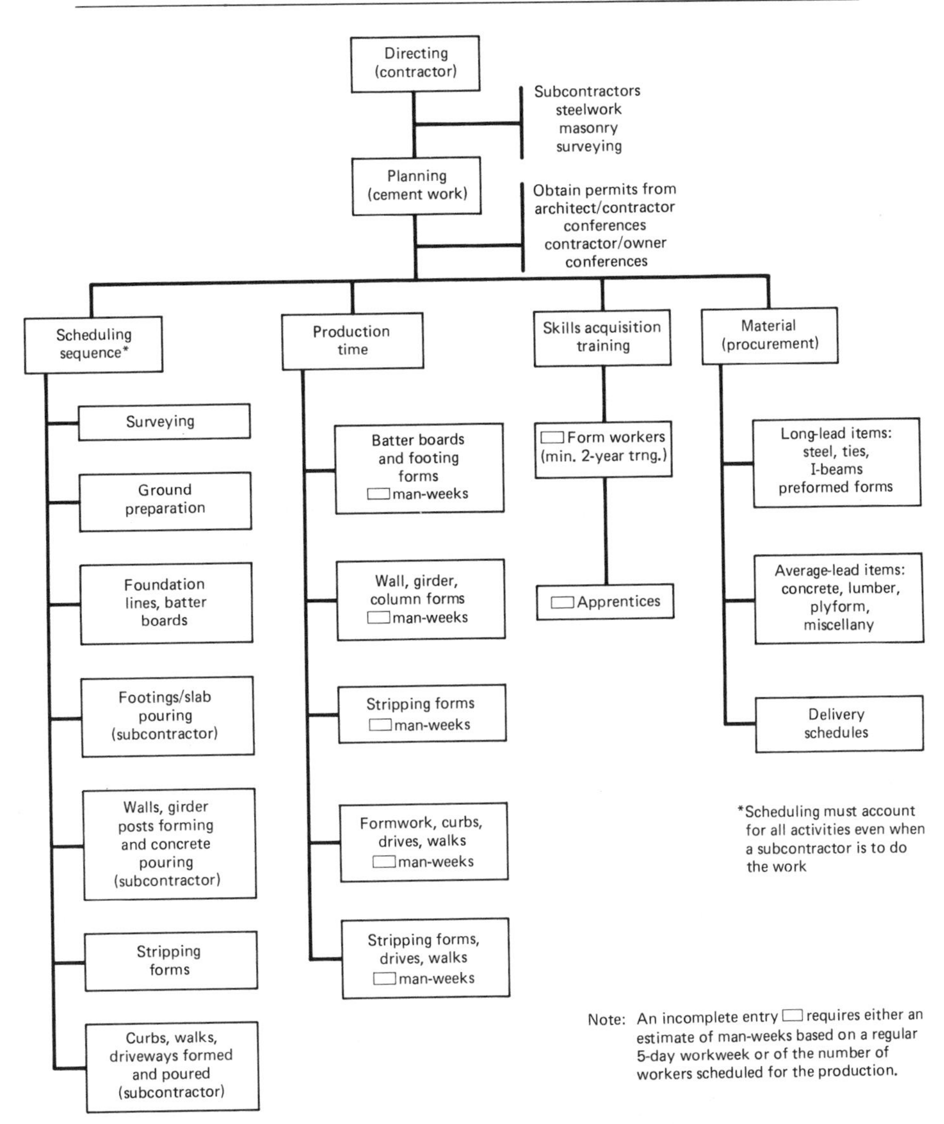

Figure 1-11 Work-flow structure of concrete formwork activities.

this example the contractor "subs" steelwork, masonry, and surveying and arranges for conferences with the "subs." He then sets four parts into being regarding this phase:

1. A scheduling sequence which must account for *all* activities even when a subcontractor performs them.
2. A schedule of production time usually listing the major elements of the work but accounting for minor elements as well. Time is allocated to each.
3. Next, he seeks skills to complete the variety of tasks and decides if apprentices can or should be employed.
4. Finally, he plans for the procurement of long- and average-lead-time items and obtains delivery schedules.

With the four elements of the plan completed, the contractor or his scheduler is ready to prepare the time-line plan. This is a plan listing job elements in the left (leader column) and times in the remaining columns. Dates and times and types of personnel skills are listed.

Dates

July 1 2 3 4 5 6 ... 15

1. Surveying
2. Ground prep
3. Foundation, batter boards, lines
4. Plumbing below slab
5. Footings/slab

This brief example shows that time-line planning/scheduling can be accomplished at any level of detail needed and can be done for a variety of purposes:

1. Work crew schedules
2. Subcontractor schedules
3. Labor hiring schedules
4. Material ordering and delivery schedules
5. Inspection schedules

Multijob Work Flow

Now let us examine briefly how a *multijob* work-flow structure would appear. Suppose that we are to build 12 condominiums or apartments. We could easily start by defining the seven phases involved in the single-job concept discussed earlier. Our plan would look as shown in Figure 1-12.

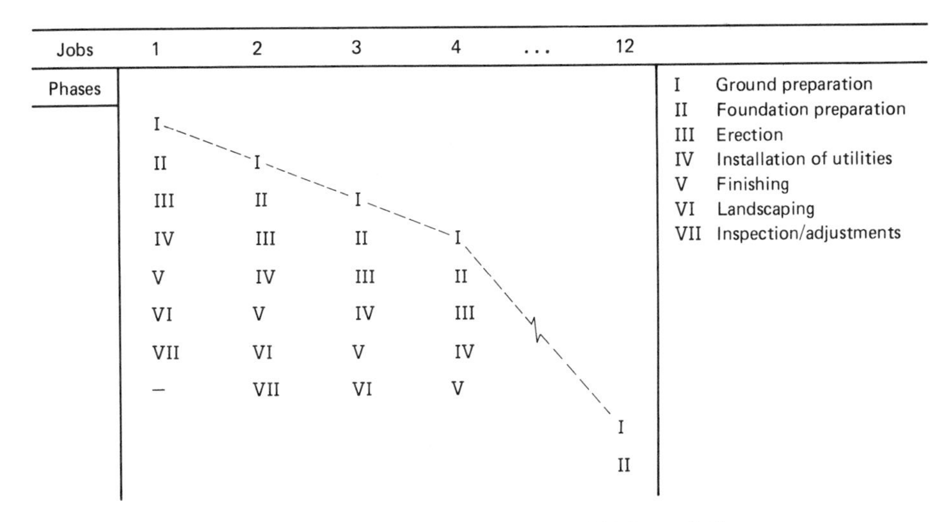

Figure 1-12 Seven phases under multijob work flow.

Then, of course, we would consider what could be consolidated in ordering supplies, timing subcontractors, and utilizing work crews to reduce waste and take advantage of discounts and other dollar savings.

You can see that this sort of planning, although vital to efficient operations, becomes unwieldly as the job size increases. Therefore, many contractors are now using computer-generated output reports for scheduling. There are a variety of software programs usable on many types of microcomputers that do scheduling and work-flow work. All you do is feed in your specific requirements and the computer does the rest. Several names are applied to these programs:

1. PERT (Program Evaluation and Review Technique), established for the Navy by IBM.
2. CPM (Critical Path Method), used both on large-scale construction projects and to create efficiency for small jobs that may increase profits and bonus opportunities.
3. "Project Manager," software.
4. "Manufacturing Inventory Control," software.
5. "Job Scheduling," software.
6. "The Estimator, or Estimating," software.

REFERENCES AND SOURCES

1. Technical Publications Department, *A Dynamic Project Planning and Control Method* (New York: IBM, n.d.); includes PERT/CPM
2. Software for desktop computers.
3. Desktop computer manufacturers (i.e., Apple, Radio Shack, IBM, Victor, and Osborne)

COMMUNICATIONS AND MANAGEMENT

Basic Steps

Basic steps to communication, as shown in Figure 1-13, include the following: source, symbols, receiver, feedback, and confirmation.

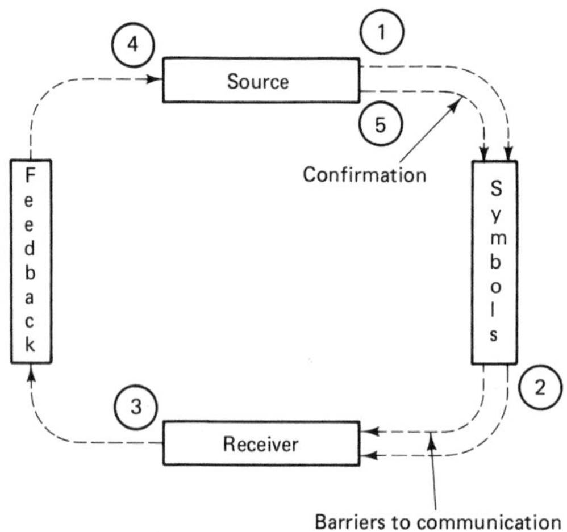

Figure 1-13 Basic steps to communication.

Source. The source is the originator, who creates the message that is to be sent to, for example, staff members to obtain financial reports or summarize data related to a job. Or the source may be a sender, who sends a message to a foreman, for example, to finish a job before Friday. Sources may be written, audio, or video messages. The variety of sources include:

1. Oral messages
   a. Managers, supervisors, and specialists directing operations
   b. Staff conveying reports
   c. Trainers instructing apprentices
   d. Foreman and architect or journeyman discussing project specifications or problems
   e. Subcontractors scheduling their operations with the general contractor
   f. Contractor dealing with bankers, inspectors, buyers, and sellers
   g. Salespersons negotiating for new projects
2. Written messages
   a. Building plan specifications
   b. Blueprints

   c. Production plans
   d. Accounting budgets and reports
   e. Field messages
   f. Company policy and construction standards
   g. Posters on a variety of subjects
   h. Paychecks
   i. Termination notices
   j. Textbooks on construction fundamentals and references
3. Video messages
   a. Training films
   b. Demonstrations to contractors, workers, trainers, etc.
   c. Television productions that teach or illustrate construction or construction-related material

Symbols. The *symbols* used in communication are varied and generally are either concrete or abstract examples. All people who act as senders must be absolutely sure that the symbol selected means the same thing to all receivers. Some symbols include:

1. Letters, words
2. Sentences, paragraphs
3. Measurement symbols: ft, in., yd, in.^{2}
4. Drawing symbols and scales
5. Plumbing, heating, and electrical symbols
6. Engineering symbols
7. Detail drawings
8. Elevation drawings
9. Floor plans
10. Safety symbols, danger symbols
11. Colored symbols

Receiver. The *receiver* is the person or persons to whom the message is directed. The source designs the symbols carefully so that the receiver will understand the message. Each of the examples of sources could well be the receiver in many instances.

Feedback. The *feedback* is the receiver forming a message with a set of symbols. He or she then transmits the message to the source. The methods of feedback include.

1. Audio
2. Gestures
3. Written
4. Compliance

Confirmation. *Confirmation* completes the cycle. The source transmits a new message that either affirms that the receiver correctly understands or has acted on the message, or disapproves of the feedback as incorrect and retransmits the original message.

Communication Paths

There are several methods of communicating that produce a variety of desired results. Each has both good and bad qualities, so care must be used when selecting one or more for transmitting or receiving messages.

I or Straightline Method. The I or straight-line method of communication is extremely weak when the communication path involves more than one receiver. An example is provided in Figure 1-14.

When using the I or straight-line method, each receiver hears what he thinks is the original message, but as you see, in this case there will be a minimum of 5 idle man-hours, if not more.

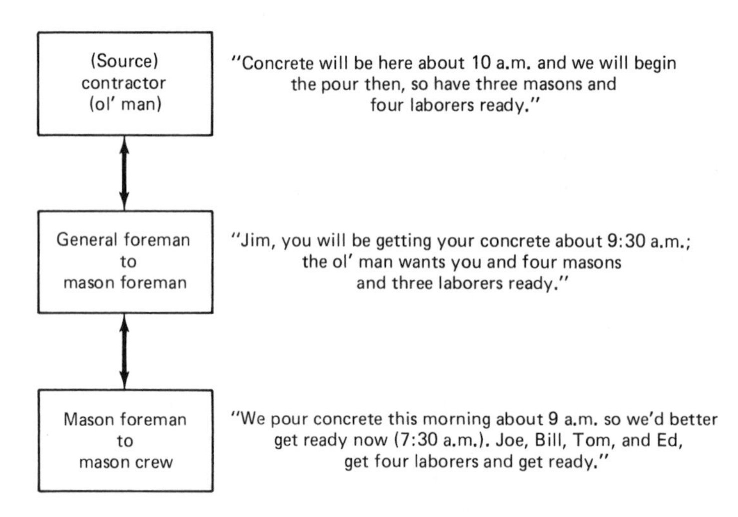

Figure 1-14 I or straight-line communication method.

Y or Inverted-Y Method. The Y method or inverted-Y method is better than the I method when oral communication is used. Using our construction project, we see in Figure 1-15 that fewer errors are made.

Notice that using the Y method to communicate results in less opportunity to distort the facts.

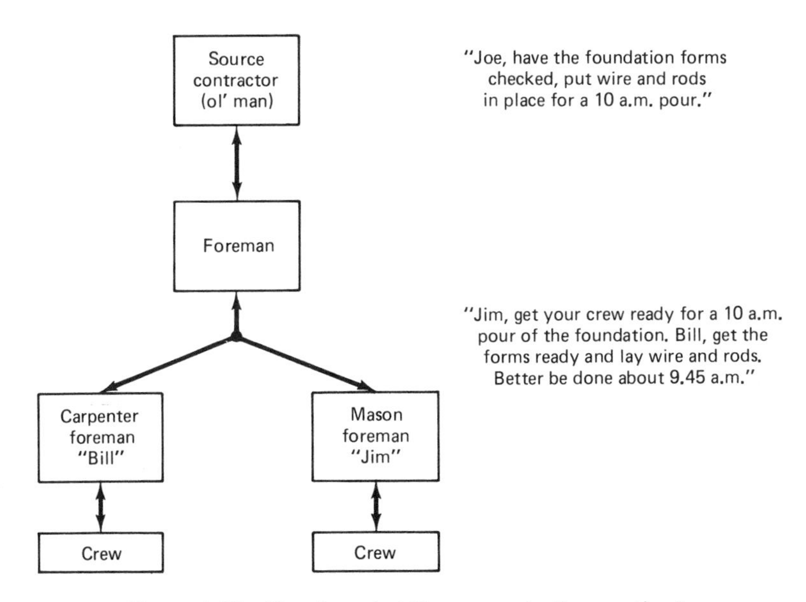

Figure 1-15 Y or inverted-Y communication method.

Spoke or Radial Method. The source occupies the *hub* of the spoke or radial method, as shown in Figure 1-16. The source is now capable of communicating with each *spoke* or *radial* and receiving from each.

Both the strength and weakness of this method is that there is no communication among spokes. If absolute control is desired, the system is the strongest one possible. However, if the subcontractor wishes to communicate with the architect, he or she must communicate to the source, who retransmits to the architect, and the response traverses a reverse path.

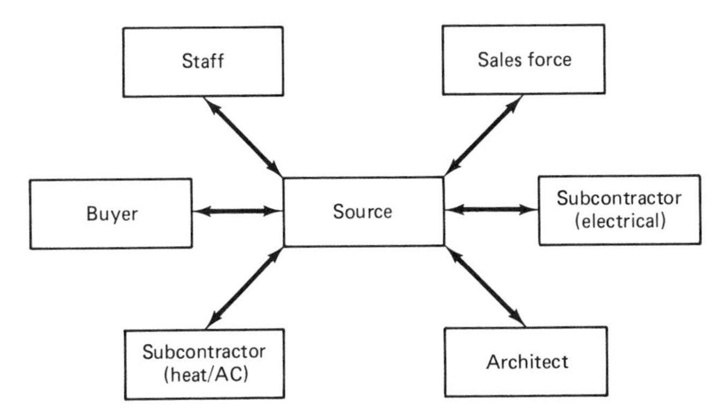

Figure 1-16 Spoke or radial communication method.

Interlace Method. The *interlace* method, illustrated in Figure 1-17, offers the widest possible paths of communication. Everyone in the communication stream can communicate directly.

The strength to communicate across channels may become a weakness if, for example, a subcontractor uses all but one path, the foreman. By omitting the foreman, the job progress may become faulty and costly; idle time could increase.

From this brief examination of communication we have learned that the wheel of communication must be complete if accurate compliance with decisions is to be carried out. Well-designed communication means profit. Poor communication means waste, idle time, and possible injury as well as a loss of profit.

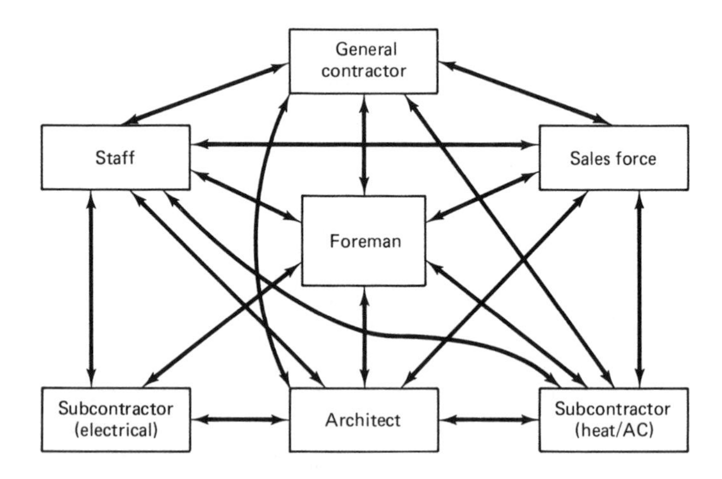

Figure 1-17 Interlace communication method.

SOURCES

1. Educational sources
   a. Institutions
      (1) Junior colleges
      (2) Business schools
      (3) Senior colleges
      (4) Postgraduate studies
   b. Subjects
      (1) Business administration
      (2) Personnel management

(3) Management of the firm
(4) General management principles
(5) Labor relations
(6) Economics of the firm

2. Some publishers on the subject of management
   a. Prentice-Hall, Inc., Englewood Cliffs, NJ 07632
   b. Reston Publishing Co., Inc., 11480 Sunset Hills Road, Reston, VA 22090
   c. John Wiley & Sons, Inc., 605 Third Avenue, New York, NY 10158
   d. McGraw-Hill Book Company, 1221 Avenue of the Americas, New York, NY 10020
   e. Harper & Row, Publishers, Inc., 10 East 53rd Street, New York, NY 10022

2

FINANCIAL AND ACCOUNTING MANAGEMENT AND ECONOMIC PLANNING

QUICK REFERENCES

As you can see from the list above, a variety of subjects are included in financial and accounting management and economic planning. All business managers must be concerned constantly with these elements. Business managers add significantly to an enterprise's ability to maintain a share of the market, direct its growth, and provide constant source of employment for its employees. To do this successfully in today's business world, given its economic conditions (recessions, low growth and periods of growth, fluctuating interest rates, etc.), modern contractors and managers are relying more and more on objective methods of decision making versus subjective methods or "gut feelings." However, experience and gut feelings still play a significant role in both financial and accounting management and economic planning.

DECISION MAKING UNDER RISK AND/OR UNCERTAINTY

If we as contractors and managers were able to control the timely arrival of supplies, have just the exact number of workers with exactly the right skills, and were able to either control the weather or pick a period of x days needed for job completion during fair weather, there would be absolute *certainty*. But the real world is not likely to combine these and other factors that we identify with certainty. Therefore, we always operate in the area of risk and uncertainty in our decision-making processes. If we use the proper decision factors (probabilities, weather forecasts, economic indicators, financial ratios, suppliers' schedules, etc.), we can limit risk and uncertainty. This leads to successful operations and company profit and growth (firm's goals).

Risk

Risk involves the objective approach to decision making. A situation involving risk is defined as follows: All the possible outcomes from a decision are known and each has definite probabilities, also known by the decision maker. For example, a contractor must decide between building a new house during a 3-month period or contracting to refurbish an apartment complex during the same period. The work force has the skills to do either project. Materials for each are readily available. The risk is associated primarily with the weather. If the weather is good, the contractor could easily build the house and make a good profit. However, if the weather turns bad, he or she could be penalized for failing to complete the job on time. On the other hand, if the contractor opts for the refurbishing project, weather is not a factor, but profits are expected to be lower. Which project should the contractor undertake?

Since weather is not a factor in the apartment project, the probability of success (all other factors considered equal) is 100% ($P = 1$). But two considerations affect the probabilities assigned to the first project. The first is *a priori*; this is where mathematical or physical principles are known or can be accurately assessed. In our example, weather statistics secured from the regional or local weather bureau would provide the past several years' data on number of sunny days, rainfall, and temperatures. From these, probabilities can easily be established (i.e., 30% rainy days = $P_{(r)} = 0.3$; 70% sunny days = $P_{(s)} = 0.7$). Now the contractor can better assess the chance to earn the larger profit.

The contractor might also have another type of risk situation, called *a posteriori*, where he or she has in the files or otherwise available statistics of similar projects built under similar conditions. These past experiences are factual and can be used in assigning the probabilities of risk.

Thus we understand that *risk* is very real and it is an objective method of appraising a tentative project's value. In a contract, *uncertainty is subjective.*

Uncertainty

The reason uncertainty is considered subjective is inherent in its definition. Whereas situations of risk can have probabilities applied based on fact, outcomes based on uncertainty have no factual background. These may include gut feelings, general opinion, "someone says," or earlier experiences in similar situations but with different variables. Therefore, *best-guess probabilities* or *equally likely probabilities* are chosen subjectively. If the values chosen end up being comparatively accurate, great—the project is successful. However, if they are off track—the project may well fail to produce a profit.

Given our sample problems, if any variables were subject to uncertainty, the likelihood of chosing that project decreases. Which brings another idea forward. The buyer who contracts the building or other project may be in much the same position as the contractor. So he or she may require that a bond be posted by the contractor to reduce uncertainty and improve or lessen the risk. The bond acts like an insurance policy. Contractors also carry various forms of insurance on their own behalf to lessen the likelihood of these factors.

The Decision-Making Process

Where a contractor or manager must evaluate the desirability of bidding or not bidding on a project that includes risk and/or uncertainty, he or she

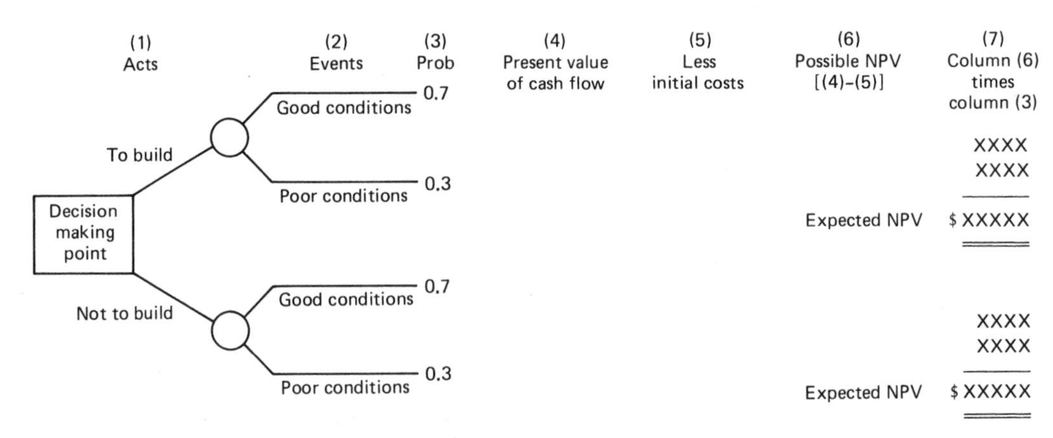

Figure 2-1 Decision tree.

must apply some decision-making process. For those who own desktop computers, software programs are available that make work easy. For those who do manual evaluations, a hand-held calculator really simplifies the variety of mathematics involved.

One decision process involves three steps: These are sometimes included in *decision trees*. In making a decision tree one lists (1) the *acts* or possible courses of action and (2) the *events* or outcomes from each act, then (3) assigns probabilities to each event (Figure 2-1). Then the evaluator computes the various branches and selects the option that produces the best opportunity to earn a profit.

Definitions of Selection Criteria

Another decision-making process employs a variety of techniques that lead to objectively developed alternatives. They include:

1. Expected value criterion
2. Maximin criterion
3. Maximax criterion
4. Standard deviation and coefficient of variation

Expected Value Criterion. The expected value of an event is the payoff should that event occur, multiplied by the probability that the event will occur.

$$\text{EV} = \sum_{i=1} R_i P_i \tag{2-1}$$

where R = return or payoff of each event
P = probability of each event
i = 1, 2, or number of events

Maximin Criterion. This is a one-shot decision that reflects a conservative or pessimistic approach. Given the results of an evaluation, the decision maker selects the maximum of the minimum payoffs.

Maximax Criterion. This is a one-shot decision that reflects an optimistic or best opportunity approach. Given the results of an evaluation, the decision maker selects the best payoff of all payoffs.

Standard Deviation. The standard deviation of a distribution around its *mean* provides the decision maker with a measure of relative risk. The possible payoffs fall around the mean (expected value). The larger the standard deviation, the greater the risk.

$$\sigma = \sqrt{(X_i - X)^2 P_i} \tag{2-2}$$

where X = expected value
X_i = outcomes
P_i = probability of each occurrence

Coefficient of Variation. The decision maker would select the project that produces the *smallest* coefficient of variation, which is found by dividing the standard deviation (SD) by the expected value (EV).

$$\text{Coefficient of variation} = \frac{\text{SD}}{\text{EV}} \tag{2-3}$$

Decision making under risk and/or uncertainty is the real-world situation for all contractors and managers in the construction industry. The process of assessing risk and/or uncertainty is one where risk produces more favorable results because of its objective approach. Not only can different projects be assessed, but the decision process also applies to:

1. Buying or leasing equipment
2. Contracting labor or hiring workers
3. Subcontracting or doing with firm resources
4. Borrowing money to finance the project or finance from company resources

Finally, remember that computer software programs are available that take the tedium out of the task and provide useful data for decision making.

REFERENCES

1. Douglas, Evan J., *Managerial Economics* (Englewood Cliffs, N.J.: Prentice-Hall, 1979).
2. Simon, J. L., *Applied Managerial Economics* (Englewood Cliffs, N.J.: Prentice-Hall, 1975).

FINANCIAL RATIOS

Financial ratios (see Exhibit 2-1) are especially important to contractors and managers because they provide a quick indicator of certain aspects of the firm, including its ability to make profit, satisfy stockholders, secure added capital, and define how well customers are paying their accounts. By and of themselves each ratio does little to explain; however, when used with similar earlier ratios or industry standard ratios, one can make serious judgments. If questions arise when making these comparisons, a study is made to unearth the cause of the variance.

A publication widely used by the lending institutions and businessmen, including contracting firms, is the *Annual Statement Studies* developed and published by Robert Morris Associates. Though the publication covers the

EXHIBIT 2-1, FINANCIAL RATIOS—CONSTRUCTION INDUSTRY[1, 2]

TYPE	MEDIAN RANGE
Current	1.0- 1.8
Receivables/Payables	1.0- 1.6
Revenues/Receivables	4.8-12.7
Revenues/Working Capital	9.9-17.2[3]
EBIT/Interest	1.8- 4.5
Cash Flow/Curr. Mat. L/T/D	1.6- 3.8
Fixed/Worth	0.2- 2.2
Debt/Worth	1.3- 2.6
%Profit Before Taxes/Tangible Net Worth	9.6-24.0[3]
%Profit Before Taxes/Total Assets	3.4-13.6
%Depr., Dep., Amort./Revenues	0.7- 4.9[3]
%Lease & Rental Exp./Revenues	0.4- 1.9
%Officer's Comp./Revenues	2.6- 6.8

[1] The following types of contractors are included: concrete work, electrical work, excavation and foundation work, floor laying and other floor work, commercial construction, general building—single family houses and residential buildings, heavy construction, except highways and streets, highway and street construction, except elevated highways, masonry, stone setting and other stonework, oil and gas well drilling, oil well servicing, painting, paper hanging and decorating, plastering, drywall, accoustical and insulation work, plumbing, heating (except electrical), and air conditioning, roofing and sheetmetal work, and water, sewer, pipelines, communication and power lines.

[2] Ratios are provided from the All Sizes of Contractors column.

[3] Oil and gas drilling and servicing contractors not included.

SOURCE: Robert Morris Associates, *1982, Annual Statement Studies,* pp. 353-61, (Phila., Pa.). Data reprinted and compiled by permission of Robert Morris Associates, copyrighted 1982. See disclaimer on page 68.

full spectrum of business PART IV is devoted to *Construction Contractors.* Within Part IV

> A significant feature of this edition of the Studies is the revised format of this section. Following the introduction of the new contractor audit guide by the American Institute of Certified Public Accountants, our format has been modified to make it more compatible with the industry's present accounting methods. In this section, size categories are determined by contract revenues, instead of total assets, with no upper limit placed on revenue size. Only companies reporting on a percentage-of-completion basis have been included in our sample.[1]

For each type of construction contractor included in the charts the various Assets, Liabilities and Income data entries are listed along with their

[1] *Annual Statement Studies* (Phila.: Robert Morris Associates, copyright 1982), p. 4.

respective percentage of totals. For example, the *gross profit* of contractors, involved with concrete work extensively should be about 22.7% of gross contractor revenues; whereas, General Contractors should have gross profits of about 19.9% (based on 1981 data). Percentages are provided on the following accounting summaries:

Assets. Cash and Equivalents, A/R-Progress Billings, A/R-Current Retention, Inventory, Costs & Est. Earnings in Excess of Billings and all other Current, Total Current, Fixed Assets (Net), Joint Ventures & Investments, Intangibles (Net), All other Non-Current, Total.

Liabilities. Notes Payable-Short Term, A/P Trade, A/P Retention, Billings in excess of Costs & Est., Earnings, Accrued Expenses, Curr., Mat-L/T/D, All Other Current, Total Current, Long Term Debt, All other Non-Current, Net Worth, Total Liabilities & Net Worth.

Income Data. Contract Revenues, Cost of Revenues, Gross Profit, Operating Expenses, Operating Profit, All Other Expenses (Net), Profit Before Taxes.[2]

Before providing Exhibit 2-1 Financial ratios, one more important factor needs to be identified. The ratios provided in the Annual Statement Studies are the best judgment of the authors and are provided with three quantities: *Median, Upper and Lower Quartile.* The Quartiles are a figure that falls half way between the median and highest and median and lowest points.[3] Therefore, lenders and contractors are provided a range that can more readily be applied to local situations and economic conditional changes.

Working Capital Ratios

Working capital ratios identify generally how well the firm is able to pay its short-term debts.

$$\text{Current ratio} = \frac{\text{current assets}}{\text{current liabilities}} \tag{2-4}$$

Data taken from balance sheet.

Rule of thumb: ratio should be 2:1 or greater.

$$\text{Acid-test ratio} = \frac{\text{cash} + \text{marketable securities} + \text{accts receivable}}{\text{current liabilities}} \tag{2-5}$$

Inventory is excluded.

Prepaid expenses are excluded.

Rule of thumb: ratio should be 1:1 or slightly greater.

[2]Ibid., pp. 353, 355.

[3]Ibid., p. 7.

Accounts Receivable Ratios

The *accounts receivable turnover ratio* shows how well accounts are being paid by those who have credit with the firm. This leads to defining the average days to collection from creditors.

$$\text{Accts receivable turnover} = \frac{\text{sales (in dollars)}}{\text{average accts receivable balance}} \tag{2-6}$$

$$\text{Average collection period (days)} = \frac{365}{\text{average receivable turnover}} \tag{2-7}$$

Inventory Ratios

The *inventory turnover ratios* measure how many times a firm's inventory has been sold during a year. This ratio may not apply to small companies that do not maintain inventories. However, corporations that buy in carload lots would definitely need this information. Storage warehousing of inventories costs considerable money (buildings, yards, warehousing, insurance, security, inventory control, etc).

$$\text{Inventory turnover} = \frac{\text{costs of goods sold + times}}{\text{average inventory (dollars)}} \tag{2-8}$$

$$\text{Number of days to sell inventory} = \frac{365}{\text{inventory turnover}} \tag{2-9}$$

Times-Interest-Earned Ratio

The *times-interest-earned ratio* provides a measure of the company's operation to pay interest to long-term notes held by its lenders.

Times interest earned

$$= \frac{\text{earnings before interest expense + income tax}}{\text{interest expense}} \tag{2-10}$$

Debt/Equity Ratio

The *debt/equity ratio* measures the portion of assets being provided by creditors (total liabilities) against the portion of assets being provided by owners (stockholders) (equity). Creditors like a low debt/equity ratio, owners prefer higher ratios.

$$\text{Debt/equity ratio} = \frac{\text{total liabilities}}{\text{owners' (stockholders') equity}} \tag{2-11}$$

Stockholders' Earnings Ratios

Earnings per share concern people who expect to derive income from distribution of profits.

$$\text{Earnings per share} = \frac{\text{net income} - \text{preferred dividends}}{\text{common shares outstanding}} \tag{2-12}$$

The *return on common stockholders' equity ratio* is a percentage that indicates the rate of return a company can generate based upon common stockholders' equity in the company.

Return on common stockholders' equity

$$= \frac{\text{net income} - \text{preferred dividends}}{\text{average common stockholders' equity (preferred stock excluded)}} \tag{2-13}$$

The *price/earnings (P/E) ratio* indicates the relationship between the market price of a stock and the current earnings per share for that stock. Growth companies usually have high P/E ratios.

$$\text{P/E ratio} = \frac{\text{market price of common stock}}{\text{earnings per share of common stock}} \tag{2-14}$$

The *dividend payout ratio* is a measure of the percentage of current earnings being paid out in dividends. Investors seeking ordinary income want high percentages; investors who seek capital gains want low percentages.

$$\text{Dividend payout ratio} = \frac{\text{dividends per share}}{\text{earnings per share}} \tag{2-15}$$

The *dividend yield ratio* provides stockholders with a yield that would be lost or sacrificed if the investor sells the stock at its current market value (opportunity cost).

$$\text{Dividend yield ratio} = \frac{\text{dividends per share}}{\text{market price per share}} \tag{2-16}$$

Return on Total Assets

The *return on total assets ratio* provides a percentage value that measures how well the assets of a firm are employed in the goals of the firm or how well it is performing.

$$\text{Return on total assets} = \frac{\text{net income} + \text{interest expense}}{\text{average total asset}} \tag{2-17}$$

Note: Average total assets are obtained by *adding* beginning-of-year assets to end-of-year assets, then dividing by 2.

Book Value per Share

The *book value per share* is a measure that aids in assessing the company's well-being.

$$\text{Book value per share} = \frac{\text{stockholders' equity preferred stock}}{\text{number of common shares outstanding}} \tag{2-18}$$

Note: Book values are compared with market value, but usually are used to reflect past company actions and efforts and may be useful in defining a company's worth and setting the price of common stock.

Operating Leverage

Operating leverage measures the change in net income for a change in revenue. It is useful in determining the impact that a contract will have on the firm.

$$\text{Operating leverage} = \frac{\text{contribution margin}}{\text{net income}} \tag{2-19}$$

SOURCES

1. Financial statements (Chapter 3)
2. Audit reports (Chapter 3)
3. Managerial accounting text books

Final note: Whether we are operating a sole proprietorship, partnership, or corporation, some, if not all, of these financial ratios are very important to our year-to-year understanding of how well we are performing. By careful evaluation of the various ratios, contractors can take steps to meet future needs for:

1. Growth
2. High profit margins
3. Greater efficiency
4. Debt levels
5. Securing capital
6. Meeting other needs of the firm

FINANCIAL FORECASTING AND TERMS

Probably the most difficult part of management is the forecasting of financial strategies because so much of the construction industry is involved in contracts that are of short duration (less than a year). Second, construction projects are contracted for by those needing buildings, improvements, and larger construction projects when money supplies are most favorable or the current economic climate of the country indicates growth. Managers may be faced with short-term, long-term, or a combination of short- and long-term financial decisions affecting their company. Therefore, we shall identify some of the more important tools available in making decisions.

The Cash Flow Problem and Budgets

Short-term cash flow is always a problem for construction managers. Salaries must be met weekly, accounts payable and debts must be paid to creditors, insurance premiums and taxes must be paid periodically, and inventories and lease payments must be made. Many companies do not have adequate cash reserves to meet these and other cash requirements on a timely schedule, and thus find themselves in crisis situations. The use of a forecast cash flow budget helps to locate deficit cash flow periods and allows managers time to secure short-term credit.

CASH FLOW BUDGET 19xx

Income:	JAN	FEB	MAR	APR	MAY	JUN	JUL	AUG	SEP	OCT	NOV	DEC
Bal brot fwd	xx	xx	xx	xx	xx	xx	xx	xx	xx	xx	xx	xx
(1) Source	xx	xx	xx	xx	xx	xx	xx	xx	xx	xx	xx	xx
(2) Source			xx									xx
Total income	xx	xx	xxx	xx	xx	xx	xx	xx	xx	xx	xx	xxx
Expenses:												
(1) Source	xx	xx	xxx	xx								xx
(2) Source	xx			xx				xx				xx
(3) Source		xx	xx	xx	xx	xx	xx	xx	xx	xx	xx	
Total expense	xx	xx	xxx	xxx	xx	xx	xx	xxx	xx	xx	xx	xx
Net income (loss)	xx	xx	xx	(xx)	xx	xx	xx	(x)	xx	xx	xx	xx
Short-term loans				xx				xx				

This type of cash flow budget is easily made and usually contains data based on concrete, empirical information.

The difficulty for any company involved in a variety of mutually exclusive construction projects with similar or varying length makes this budget difficult but not impossible to construct. Here, again, a desktop microcomputer is a valuable aid.

Each source of income should be identified by job number and/or title. If payment schedules are stipulated in the contract as a function of percent of job completion, the manager must translate these percentages into calender days or months to include them in the budget. Short-term loans may also be indicated as sources of income, *but usually are not.* They are listed below net income (under expense) and then are added to the balance brought forward in the following month under income.

Each expense needs to be forecasted as accurately as possible. Understating them creates a favorable cash picture when one does not exist. Overstating expenses causes a lower than actual need for cash as well as short-term notes that are not necessary.

Several other considerations need to be made when forecasting cash flow. The adequacy of estimation for future jobs usually involves changes in productivity in that different geographical on-site and locale conditions will exist, numbers and kinds of workers will differ, and different machines may be used. Second, pricing for these factors as well as material inventories may differ. To account for these, job estimate data modified appropriately will be used for expenses and are then consolidated with other projects in the time period prior to adding the data to the budget.

Inventory Cost Factors

Inventories of materials dedicated to present and future projects are a company's asset. They can be used to secure short-term loans to meet cash flow problems; for example, inventories must (1) be used quickly to generate company profit and operating funds, and (2) must be kept as low as possible to avoid costs incurred.

1. Costs associated with inventories:
   a. Storage at x dollars per square foot
   b. Time or number of days or months inventory turns over
   c. Insurance premiums
   d. Security costs
   e. Inventory accounting costs
2. Economic order quantity:

$$\text{EOQ} = \frac{\sqrt{2 \times \text{annual materials required} \times \text{cost per order placed}}}{\text{storage cost per unit}} \qquad (2\text{-}20)$$

Defined: EOQ is the optimum quantity of specified inventoriable materials to be ordered at one time. Here is a method whereby the manager can save considerable money by (1) placing large enough orders to obtain discounts and yet hold warehouse (inventory) costs to a minimum. Again a computer program can easily provide a solution.

Example: A company expects to build 12 homes in one year. Each home uses 9 window units (12 × 9 = 108 units).
Question: What are the EOQ costs?

Number of orders	Number of units per order	Average units in inventory	Order cost at $12 each order	Storage cost at $1 per unit	Total cost
1	108	54	$12	$54	$68
2	54	27	24	27	54
3	36	18	36	18	54
4	27	9	48	9	57

Note:

1. By using the sample above, a manager could select either two or three orders per year, since the total costs are equal.
2. Using the EOQ formula, the best option would be order twice a year.

We have illustrated window units, but each type of building material can be handled the same way *if* the company expects to have recurring needs for like construction items throughout the year. EOQ does not work very well on short-term (3-month, 4-month) projects.

Increasing Owners' Equity

Owners' equity (capital) in the company can take many forms. These include but are not limited to:

1. Cash placed in the company by the owners
2. Capital stocks issued by the company
3. Equity on other property assigned to the company
4. Equipment
5. Real and personal property
6. Retained earnings held for future investment

Securing Capital

Securing capital for continued operation, expansion, and other reasons that would lead the company to achieving its goals is a constant part of the job of all managers. Following are some techniques used to increase capital:

1. Issue additional stock. This technique increases capital but dilutes current stockholders' value in dividends since more people split the profit dividends.
2. Issue preferred stock. These act like bonds in that preferred stockholders have first claim on net assets in the case of default.
3. Pledge future earnings as security for long-term notes and sometimes for short-term notes.
4. Obtain additional partners (if a partnership).
5. Convert the firm from a sole proprietorship to a partnership or corporation.
6. Make a profit on every contract and retain earnings within the company.
7. Issue long-term bonds. This is a company debt, but the thinking is that the increase in capital will be invested to produce profit in excess of bond repayment.
8. Establish depreciation schedules using best depreciation technique (straight-line 150% declining balance, 175% declining balance, double declining balance).

Capital Budget Decision Making

Capital budget decision making that can increase capital and fall within the operating segment of the company includes:

1. Reducing labor-intensive operations through mass production and increased use of assemblies and subassemblies
2. Purchasing new, more efficient equipment
3. Deciding on lease or purchase of equipment and plants
4. Making plants more efficient
5. Careful screening of construction projects; accepting those that have the greatest internal rate of return (IRR)
6. Reducing loss caused by on-site waste and pilferage
7. Treating exchange or sale of capital assets as net short-term capital gains (or loss) and net long-term capital gains (or loss) that will affect the tax liability of the company
8. Investment credit on new purchases of equipment and certain real property

Present Value (Discounting) Analysis

Since most construction companies in the industry operate under contract, the use of present value analysis is attractive to bidding on contracts and determining future cash flows. Since many companies need to borrow to finance their contracts, receive cash flows in future periods, and must maintain internal rates of return (IRR) satisfactory to remain in business, the PV analysis concept is very important.

$$\text{Present value (PV)} = \frac{R}{(1 + r)^n} \tag{2-21}$$

where R = revenue to be received in the future
r = rate of interest charged to borrow on earnings or IRR expected from project to firm's capital
n = number of years in the future the sum will be received (partial years are fractional quantities)

Computing this for each case would be laborious. Computer software programs can calculate the PV given R and r, or tables of PV such as those given at the end of this chapter make computation easier.

Net Present Value. The net present value (NPV) analysis of a project must be *positive* throughout the investment period or we should reject the investment. If the net present value of a project's inflow of income exceeds the present value of the investment, the project adds to the firm's equity.

EXAMPLE 1: Suppose that a mason contractor is considering buying a gasoline-operated mortar-mixing machine to support his operation for the next 5 years. He estimates that one less part-time laborer will be needed, with a resulting labor savings of $4000 per year. The mixer's initial cost is $11,500.

Summary:

Initial cost	$11,500
Life of project (years)	5
Annual savings in labor costs	$4,000
Desired rate of return	20%
Salvage value	0

Calculating:	Years having cash flow	Amount of cash flow	20% rate of return factor	Present value of cash flow
Initial investment	0	$—11,500	1.000	$—11,500
Annual cost saving	1-5	4,000	2.991	11,964
Net present value				$ 464

EXAMPLE 2: Suppose that we want to determine the effect on NPV when depreciation and tax savings are included.

Summary:

Depreciation (straight-line), 5 years	$2,400
Tax bracket	50%
Tax savings on depreciation	$1,200 per annum

Calculations year	Contribution	Depreciation	Tax savings	Cash flow after taxes	Discount factors	Net present value
0	$—11,500	—	—	$—11,500	1.000	$—11,500
1-5	4,000	$2,300	$1,150	5,150	2.991	15,404
						$ 3,904

$$\text{NPV} = \sum^{n} \frac{R_t}{(1+r)^n} - I_c \qquad (2\text{-}22)$$

where $t = 1$
$n = 1$
I_c = initial cost

When depreciation and the tax savings are considered, the contractor is able to operate at a lower discount rate and still earn equity if he or she wishes.

Internal Rate of Return. The internal rate of return (IRR), sometimes called the time-adjusted rate of return, is usually calculated or defined as the true interest yield produced or projected by an investment in a project or equipment over its planned life. Now the idea is to calculate the IRR to see if it meets minimum company policy. If it does, the project should be undertaken; if below company policy standards, the project should be rejected or certain factors within the project should be reevaluated. Therefore, the IRR is the rate of discount that reduces the *inflow stream* of income equal to the initial cost of the project.

$$I_c = \sum_{t=1}^{n} \frac{R_t}{(1+r)^n} \quad \text{or} \quad I_c - \sum_{t-1}^{n} \frac{R_t}{(1+r)^n} = 0 \qquad (2\text{-}23)$$

where I_c = initial cost
n = number of years

The computer really comes in handy for this type of exercise, since identifying the IRR manually requires the use of *trial* and *error* and some interpolation. However, a hand-held programmable calculator can do this job for you.

EXAMPLE: Let us use our mason's project again.

		Net present value at:			
Year	Cash flow after taxes	20%	30%	40%	35%
0	$—11,500	$—11,500	$—11,500	$—11,500	$—11,500
1-5	5,150	15,404	12,545	10,480	11,510
		$ 3,904	$ 1,045	$ 1,020	$ 10

Results: The mason's project should produce approximations of a 35% internal rate of return.

Final note: Where income streams have different amounts per year, or where depreciations are different from straight line (SL), each year must be discounted separately, then the NPV is obtained by summing the totals for each year and subtracting the initial cost. Present value tables at the chapter's end provide the values for these computations.

Profitability Index. The profitability index is the ratio of the present value (PV) of the future income stream discounted to net present values (NPVs) to the initial cost of the project.

$$\text{PI} = \sum_{n=1}^{n} \frac{\frac{R_t}{(1+r)^n}}{I_c} \qquad (2\text{-}24)$$

The decison rule is that investment in the project is desirable when the PI is *equal to or greater than unity.*

EXAMPLE:

$$\text{PI} = \frac{\$15{,}404}{11{,}500} = 1.34$$

Final note: If a project is desirable under one system of calculation, it is desirable under the other two as well. There is no discrepancy among the systems.

VALUE OF FINANCIAL STATEMENTS

Financial statements are historical reports of past company activity. They may be from the recent past (within the current operating year) or date back to the company's beginning. Each plays an important role in the management of the company. We shall identify the variety and briefly state the value of each type to the company. (*Note:* Some actual examples can be seen in Chapter 3.)

Balance Sheet and Comparative Balance Sheet

The *balance sheet* illustrates the financial position of the company as of a specific date. It lists, among other things, assets (current and other), depreciations, liabilities (current and long-term debts), and capital or equity. From its data many financial ratios can be defined, working capital is determined, and overall growth (retained earnings and/or increase in equity) is shown. The *comparative balance sheet* is used to compare the current financial position with similar earlier periods. A column for changes (±) is usually added for ease of reading.

Income Statement and Comparative Income Statement

The *income statement* provides a picture of where and how the operations of the company gained revenue and produced goods to establish gross profits. Then it lists operating expenses. Finally, the operating income (net income) before income tax is defined. Other income and expenses are also shown (e.g., rent income, interest expenses). The *comparative income statement* serves to identify changes in revenue and expenses on an item-by-item basis. Managers then evaluate the causes for these changes.

Capital Statement and Comparative Capital Statement

1. For a sole proprietor, the *capital statement* lists beginning capital, net income less withdrawals, and the increase or decrease in capital at the end of the period. It functions to show monthly, quarterly, or annually how well the company has performed.

2. For corporations, the *comparative capital statement* provides stockholders with indications of capital stock held by shareholders and retained earnings within the company. As with the balance sheet and income statements, the comparative capital statement provides data for past years against the current year, thereby allowing comparison of company growth.

Retained Earnings Statement (Corporations)

The retained earnings statement shows how net income (after taxes) is allocated to paying dividends to stockholders and how the company's capital increases or decreases. *Comparative* statements show how well managers are carrying out the goals of the company.

Cash Flow Statement

This is one of the statements of financial position that stresses changes based on cash. A well-developed statement lists *sources of cash* from operations, reduction in taxes, investments producing cash, and the issuance of stock. It will then show *uses of cash.* These uses may include purchases of equipment, land, and facilities; the retirement of bonds and payment of dividends; and the repurchase of stocks. Finally, a summary is provided that illustrates the overall change in cash balance. This statement shows where company managers made financial decisions that *should have been* and may have been directed toward company goals.

Other Statements of Changes in Financial Position

1. *Capital or retained earnings:* This statement (previously covered) shows growth or decline in the company.
2. *Fund balance:* These statements show the change (actual or expected) of account balances over a specified period (monthly, quarterly, annually). Beginning and ending balances are shown, with explanations of added incomes and deducted expenses.
3. *Revenue over expenditures: budgeted versus actual:* This statement shows managers how well they have been operating compared to what was budgeted. A column is set aside for indicating differences. This statement can summarize the various internal accounts that were budgeted.
4. *Working capital:* In addition to listing sources and uses of working capital, this statement may and should show the *changes* in the components of working capital (both current asset and current liability components). Then the manager can better understand and explain the company's financial position in order to pay its debts.
5. *Cost of goods manufactured:* This statement explains the *work in process* aspect of a company by detailing inventory changes and the use of direct labor, materials, and overhead. Those companies that maintain inventories of building materials find this form useful in accounting for inventories and those elements usually charged to each job and overhead.

Note: If you are unfamiliar with any of these statements, ask for a copy of one from your accountant or any local accounting firm. (Some examples are shown in Chapter 3.) Just about any accounting book will also have a variety of statements that you can adapt to your needs.

WORKING CAPITAL

Working capital is defined as the excess of current assets over current liabilities at a particular period in the company's operation. Recall that these same two components were used to establish the *current ratio.*

$$\text{Current ratio} = \frac{\text{current assets}}{\text{current liabilities}} \tag{2-4}$$

The function of defining working capital is to provide an indicator of how well the company can meet its obligations. Therefore, the actual amount of working capital must be evaluated in terms of the company's needs. Working capital can be determined from a well-detailed balance sheet. Comparative working capital changes can be obtained from comparative balance sheets such as the following:

	1981	1982	Increase (decrease) in working capital
Current assets			
Cash	$10,000	$10,500	$ 500
Receivables	2,000	1,500	(500)
Marketable securities	3,000	3,000	0
Inventories	6,000	7,000	1,000
Prepaid expenses	1,200	1,000	(200)
Total	$22,200	$23,000	$ 800
Current liabilities			
Accounts payable	$ 2,400	$ 1,900	$ 500
Notes payable	3,000	1,600	1,400
Income tax payable	500	500	0
Dividends payable[a]	1,000	1,100	100
Total	$ 6,900	$ 5,100	$1,800
Working capital	$15,300	$17,900	$2,600

[a]Corporations.

Sources of Working Capital

1. Decreases in noncurrent assets (sales of property, equipment, etc.)
2. Increases in noncurrent liabilities (issuance of long-term bonds and other securing of long-term debt)
3. Increase in stockholders' or owners' equity (revenue realized from operations, extraordinary income, retained earnings not allocated to dividends, withdrawals less than net income) (sole proprietorship)

Uses of Working Capital

1. Increases in noncurrent assets (purchases of equipment, buildings, etc.)
2. Decreases in noncurrent liabilities (retiring debt)
3. Decreases in stockholders' or owners' equity (use of retained earnings to purchase preferred stock, paying dividends, buying land, buildings equipment, etc.), withdrawals in excess of net income (sole proprietorship)

■ PRODUCTION COSTS

All costs that can be attributed to a job or contract are considered production costs. In some categories they may be wholly assigned, as, for example, direct labor and direct materials. In other categories only a portion of the enterprise's cost is assigned to the job or contract. These costs include overhead and fringe benefits. Under the *managerial accounting* method the terms used in computing job costs are as follows:

Production costs:

1. *Direct labor:* The actual man-hours of all workers.
2. *Direct materials:* The actual expense associated with all materials employed on the job.

Construction overhead: The portion of indirect expenses allocated to the contract, which may include those items below.

1. Indirect materials
   a. Allowance for waste
   b. Forms
   c. Scaffolds
   d. Security fences
   e. Weather protection
2. Indirect labor
   a. Supervision
   b. Material handlers
   c. Engineering services

d. Security guards
e. Maintenance time for equipment
f. Idle time
g. Insurance
h. Fringe benefits
i. Overtime premium

Job Order Costs

Production costs are frequently called *job order costs* by accountants and managers. In this section we look more closely into each term.

Direct Materials. The principal reason for accumulating data on the cost of materials used in the construction of the project is that it aids managers in defining profit and accounts for expenses. The office manager will make out several forms for tracking purpose: In Exhibit 2-2 we see a material requisition form. This form lists the job and its assigned number for tracking and is dated. Each material purchased for use on the job is cataloged as shown.

Job number 830116 Material Requisition Form 100

Job title 214 Spring Place Date Jan 16, 1983

Description	Quantity	Unit price	Total cost
2 x 4 - 8'	30	1.75	52.50
2 x 8	300 Lin. ft.	$0.45 bd. ft.	180.00
6 x 6 Wire	2000 sq. ft.	$0.20 sq. ft.	400.00
Concrete	20 yds.	$30 per cu. yd.	600.00
			$1232.50

Exhibit 2-2 Materials requisition form.

Direct Labor. By definition, direct labor is composed of the efforts of workers who are involved in the actual construction times their labor hour rates. Direct labor hours are usually accounted for on time sheets or time cards which are turned into the accountant periodically. The accountant (or in many smaller enterprises the secretary/bookkeeper) applies the individual wage rates to the worker's time to compute gross regular pay.

IDLE TIME. All too often, direct labor includes *idle time.* If at all possible, idle time should be listed under *overhead.*

Exhibit 2-3 shows a direct labor time sheet. The sheet (or card) needs to contain the names of all workers and foremen, the days of the week regular hours are worked, some place to show overtime hours and idle-time hours, and a place for workers to initial when they are absent for whatever

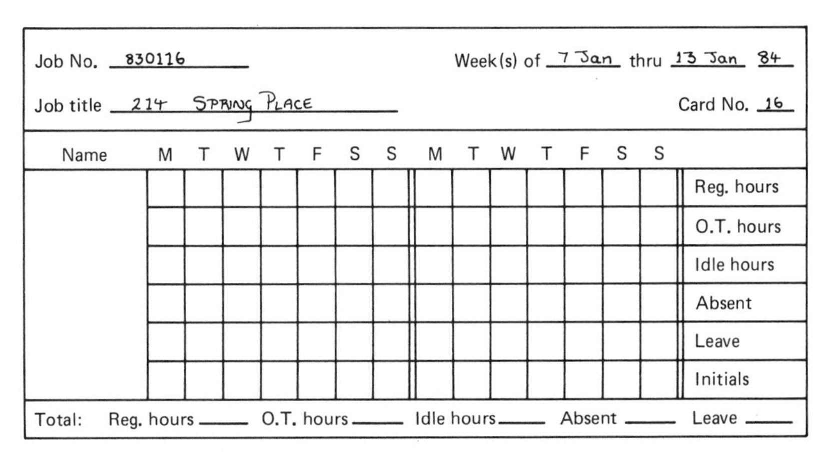

Job No. 830116 Week(s) of 7 Jan thru 13 Jan 84

Job title 214 Spring Place Card No. 16

Name	M	T	W	T	F	S	S	M	T	W	T	F	S	S	
															Reg. hours
															O.T. hours
															Idle hours
															Absent
															Leave
															Initials

Total: Reg. hours ____ O.T. hours ____ Idle hours ____ Absent ____ Leave ____

Exhibit 2-3 Time card or sheet.

reason. The sheet or card also needs to have the job number and job title for ease of filing and for use in financial accounting.

The back of the time sheet or card may be designed to compute the worker's gross pay, deductions, net pay, and the check reference (number) of the employee's paycheck.

Construction Overhead. Construction overhead, hereafter called *overhead*, are those costs that can and should be attributed to the construction project. We have listed some of them on page 49. Some need further explanation, but let us examine two ways of applying overhead.

PREDETERMINED METHOD OF APPLYING OVERHEAD. The *predetermined method* of applying overhead is computed by using empirical data from the company's prior year's operation modified by the best guess of the influences of inflation, economic conditions, and other factors, known only to the contractor. Based on these decisions, the rate is equal to

$$\frac{\text{Estimated total construction overhead (modified)}}{\text{Estimated or actual direct labor costs}} \tag{2-25}$$

$$= \text{predetermined overhead rate}$$

EXAMPLE:

Total overhead costs prior year	$25,340
Modified by estimated inflation (10%) this year	2,534
Economic forecasts for construction down 15%	(3,801)
Total estimated overhead costs	$24,072
Total direct labor hours	10,120
Modified by economic construction downturn this year (15%)	(1,518)
	8,602

$$\text{Predetermined overhead rate} = \frac{\$24{,}072 \text{ overhead costs}}{8602 \text{ direct labor hours}} = \$2.80 \text{ per DLH}$$

Suppose that our fictitious job no. 830116 is estimated to employ 1000 direct labor hours (DLH). The overhead (OH) allocated to the job would be:

Direct labor hours	1000
OH costs/DLH	× $ 2.80
Total overhead (est.)	$2800.00

Be aware, though, that these estimates must be reconciled against actual overhead costs by the accountant or bookkeeper.

ADVANTAGE AND DISADVANTAGE. The greatest advantage to the contractor in using this system is that estimating jobs becomes easier. The disadvantage is that the element of uncertainty (Chapter 2) is used and if the number or size of contracts differ drastically, severe discrepancies will exist between the rate applied and the actual overhead used.

Normalized Overhead Rates. As construction companies continue in operation year after year, *normalized overhead rates* can be used more efficiently than predetermined rates. To complete the normalized rate of overhead, the accountant would use direct labor hours for successive years (e.g., 1982, 1983, 1984) against the overhead of those years. By averaging the sums of the accounts, the normalized rate is defined.

$$\frac{\text{Average overhead costs}}{\text{Average DLH}} = \text{normalized overhead rates} \qquad (2\text{-}26)$$

EXAMPLE:

OH costs prior year 1981	$26,450
OH costs prior year 1982	25,340
OH costs projected this year 1983	26,100
Total OH costs	$77,890

$$\text{Average OH costs} = \frac{\text{total OH costs}}{3 \text{ years}} = \$25{,}963$$

DLH year 1981	9,480
DLH year 1982	10,120
DLH est. 1983	8,602
Total DLH	28,202

$$\text{Average DLH} = \frac{\text{total DLH}}{3 \text{ years}} = 9400 \text{ DLH}$$

Normalized overhead rate (NOR)

$$NOR = \frac{\$25{,}963 \text{ (average OH costs)}}{9400 \text{ (average DLH)}}$$

$$= \$2.76$$

Although the difference in computation in our examples is only 4 cents, it could be considerably more if the sizes of contracts or number of contracts differ significantly from year to year.

Fixed and Variable Overhead. Management accounting requires the differentiation of overhead into fixed overhead expenses and variable overhead expenses. Your accountant will aid in making the precise decisions as to which items of overhead are fixed and which are variable. Generally, *fixed overhead* includes the equipment needed to operate and maintain the business, the plant, and its furnishings, and related expenses and salaries of sales, supervisory, and office personnel.

Variable overhead items are those expenses directly accountable to the job and often vary with size of the job.

Now looking to our list of construction overhead and adding several more categories, we find construction overhead to include at least:

Variable overhead	Fixed overhead
Indirect materials	Plant, equipment
Indirect labor	Office and office operating expenses
Supervision	Taxes (on real estate, unemployment, etc.)
Material handlers	Insurance on real property and liability
Engineering services	Salaries of key personnel
Security guards	Office and sales salaries
Maintenance of equipment	Advertising costs
Setup time	Other sales costs
Idle time	Estimating costs
Utilities	
Fringe benefits	
Insurance (liability, theft)	
Supplies	
Equipment usage costs	
Unemployment tax	

Job Cost Sheet

Now the *job cost sheet* can be prepared. This is a summary sheet that accounts for all expenses related to the job or contract.

The job cost sheet might look as shown in Exhibit 2-4. It should contain the job number, job title, and date job began and was completed. It must also list all material requisition form data, direct labor hours, and overhead. Then a summary is compiled.

JOB COST SHEET							No. ______
Job No. ______					Date initiated ______		
Job title ______					Date completed ______		
Contract price ______							
Material		Direct Labor			Overhead		
Req. No.	Amt	Card No.	Hours	Amt	Hours	Rate	Amt

Cost Summary		Estimated Profit (Loss)	
	Amt.		
Materials	$	Revenue	$
DL	$	Less total costs	
Overhead	$ ______		______
Total costs	$	Net profit (loss)	$ ______

Exhibit 2-4 Job cost sheet.

■ CONTRIBUTION MARGIN APPROACH

As a contractor you really need to know, throughout the year, if you are making a profit, earning enough to break even, or are unable to earn enough to cover fixed expenses. To determine this information, your accountant or a computer can provide these data through a technique known as contribution margin. *Contribution margin* is defined as the revenue remaining from a contract or, on an annual basis, the total revenue that remains after variable expenses are covered that can contribute toward fixed expenses and profit. The statement looks like this:

Contract (or year's) revenue		$xxxxx
Less variable expenses		
Variable production costs	$xxxx	
Variable allocated overhead	xxxx	
Variable administrative costs	xxxx	$ xxxx
Contribution margin		$ xxxx
Less fixed costs		
Fixed production costs	xxxx	
Fixed administrative costs	xxxx	$ xxxx
Net income		$ xxxx

Uses of Contribution Margin

1. One use of the contribution margin (CM) is to determine how much of the contract's revenue will contribute to fixed expenses and profit.

EXAMPLE:

	Total	Percentage
Contract revenue	$10,000	100%
Less variable expenses	6,000	60%
Contribution margin	$ 4,000	40%
Less fixed expenses	$ 3,000	30%
Net income	$ 1,000	10%

See how easily we are able to obtain clear information.

2. A second way to use the contribution margin is for securing a contractor's building loan; this involves *operating leverage.*

EXAMPLE: Using formula (2-19),

$$\text{Operating leverage} = \frac{\text{contribution margin}}{\text{net income}}$$

$$= \frac{\$4000}{1000}$$

$$= 4$$

A banker would interpret this as follows: For every 10% increase in revenue the company is expected to earn, the percentage increase is expected to be four times, or 40%. The bank would probably be very willing to grant the loan. However, if your operating leverage were zero [CM = fixed costs (FC)], where no profit was shown, you might not obtain the needed loan.

3. A very important decision to make regarding the use of the contribution margin involves deciding on capital improvements. If the contribution margin barely covers fixed costs (operating leverage − to + 1), postponements of capital improvements would be in order. In our problem the leverage was 4 and the owners may very well decide to seek capital improvements since variable and fixed expenses are under control.
4. A contractor may feel very comfortable with a variable expense range of 50 to 65%. But he or she may well take quick and serious action if the percentage creeps above 65%. Questions that would be asked include:
   a. How efficient is the labor force?
   b. What percentage of labor time is idle time; overtime?
   c. What materials losses are there? How are materials being wasted?
   d. What overhead reductions can I make?

BREAK-EVEN ANALYSIS

Break-even analysis is used by managers, accountants, and economists to illustrate whether a company is able to (1) cover its fixed costs, (2) make not only a profit but the desired profit, and (3) bid on a contract. The break-even analysis produces the break-even point (BEP) and/or the amount of revenue needed to produce the desired profit. Now that we know something about CM, we can use the concept to determine BEP.

$$\frac{\text{Annually}}{\text{BEP}} = \frac{\text{total fixed costs (TFC)}}{\text{total CM (TCM)}} = 1 \text{ (unity)} \tag{2-27}$$

$$\frac{\text{By contract}}{\text{BEP}} = \frac{\text{allocated fixed costs to contract (AFC)}}{\text{contribution margin from contract (CCM)}} = 1 \text{ (unity)} \tag{2-28}$$

In simple terms the break-even point is always where FC = CM.

Let us use an example, illustrated in Figure 2-2. A small contract is signed to build a room onto a house for $11,352. The contractor has a small office, a truck, assorted machines and tools, and employs one carpenter on an annual basis. Past experience shows that his gross annual revenue is approximately $150,000. He expects no change this year. So this project accounts for an estimated 8% of annual earnings. He uses a normalized overhead rate of $2.75 per DLH.

Now, he has calculated during the bidding process that variable costs would account for $8500. His fixed expenses last year amounted to $12,000 and are expected to be the same this year. Will he break even, lose money, or make a profit? How much does his contribution margin help to pay fixed expenses?

Economics	Total	Percent
Revenue from job	$11,352	100
Less variable expenses	8,500	74.9
Contribution margin	$ 2,852	25.1
Less fixed costs attributable to job	960	
Net income (profit on the job) (BFIT)[a]	$ 1,892	

[a]BFIT, before income taxes.

What was his BEP for the job given the same estimates? BEP = $960.00 (where CM would be equal to fixed costs).

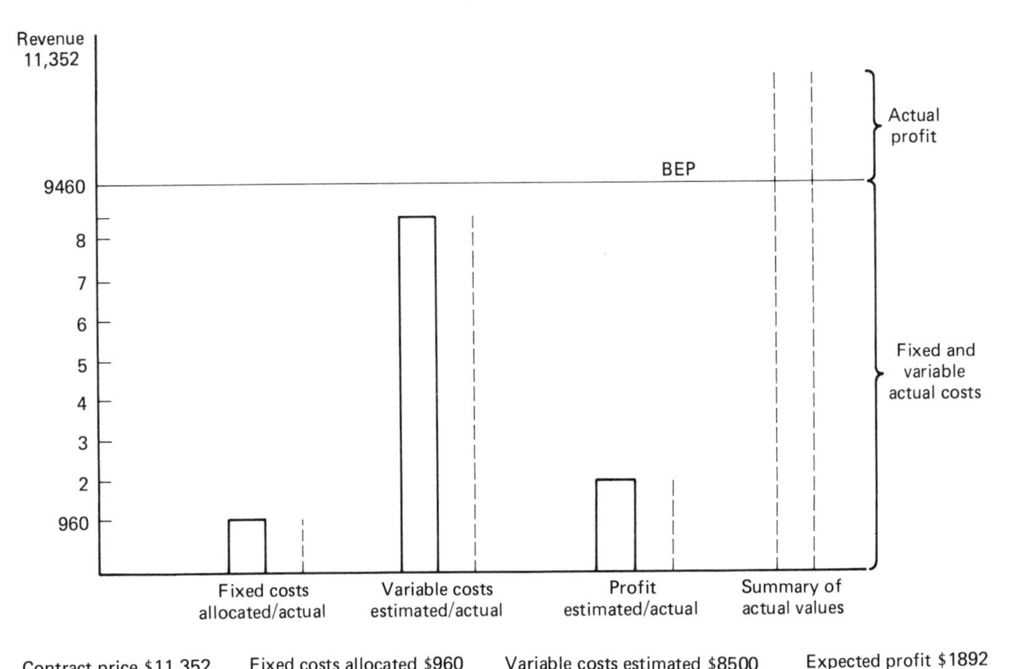

Figure 2-2 Plotting profit-center data and BEP.

Let us decide what revenue could have been the break-even point. (Incidentally, this is the figure that usually represents the lowest bid possible.)

$$\begin{aligned}\text{Price to be} &= \text{contract variable costs} + \text{allocated fixed costs}\\ &= \$8500 + \$960\\ &= \$9460 \qquad (2\text{-}29)\end{aligned}$$

BEP with Profit

Let us add one more element, profit at a specific percentage, say 20% over the total fixed cost (FC) and variable cost (VC) attributable to the job.

$$\begin{aligned}\text{Revenue} &= \text{VC} + (\text{CM} = \text{FC}) + \text{profit}\\ &= \text{VC} + (\text{CM} = \text{FC}) + 0.2[\text{VC} + (\text{CM} = \text{FC})]\\ &= \$8500 + \$960 + \$1892\\ &= \$11{,}352 \qquad (2\text{-}30)\end{aligned}$$

By using the information and formulas above and by applying overhead rates in conjunction with direct costs, a contractor can figure the bid price using very sound data simply.

Remember, though, a job is acceptable where the revenue generated causes the contribution margin (or marginal income) to equal the fixed costs attributable to the job. Any revenues in excess are income (profit from operations). A contractor may accept these types of contracts to keep crews working as well as to cover the fixed expenses for the year.

Uses of Break-Even Analysis

1. To identify the minimum bid price that produces revenue to absorb all variable costs and a designated percentage of fixed costs
2. To compare the profitability of two or more contracts that could employ the firm's resources
3. To compare the desirability of leasing or buying equipment to complete a construction project
4. To compare the desirability of hiring workers to perform or subcontracting various elements of the construction project
5. To define operating leverage

■ PROFIT PLANNING

There are several profit planning methods any contractor may elect to use with his or her company. One very desirable type is the *profit-center method*; another is the *target-net profit approach;* a third is the *cost-plus pricing and profit;* and a fourth is the *profit objective and cost-plus pricing combined.*

Profit-Center Method

A contractor who elects to use the profit-center method of profit planning assumes that *each contract is a profit center method section.* This means that each contract's revenue and costs (allocated variable and fixed expenses) are accounted for in determining if a profit is made. The accountant would determine net income by normal accounting methods, thus charging all expenses appropriated to the contract. Or the accountant may provide both: accounting records for tax purposes as just stated, as well as providing the contractor with a contribution margin approach report to keep the contractor apprised of his or her profitability status. Advantages include:

1. The contractor is aware of his or her status with each contract in force.
2. Controls with expenses are more readily set in motion.
3. Visual determination of the percentage of fixed costs is accounted for. Thus, summing the allocations gives a quick insight into annual projected fixed costs.
4. Visual determination of the percentage of profit (net income) is accounted for. The summing of profit from completed contracts gives the picture of profitability.
5. The expenses and profits can be easily charted, and ratios are constantly available.

Target Net Profit Approach

In the *target net profit approach*, the manager picks a target net profit from operating the company for a specific period of time or contract (usually time). For this we use a slight variation of equation (2-30).

$$\text{Revenue} = \frac{\text{fixed costs} + \text{net profit desired}}{\text{CM as a percentage}} \tag{2-31}$$

EXAMPLE: Suppose that a company wishes to earn a profit (before tax) of $45,000 in one year. Its fixed expenses are normalized at $25,000; direct materials and labor expenses are estimated and overhead is normalized such that the total variable expenses equal 60% of revenue (sales). What annual revenue would be the company's goal? Using equation (2-31):

$$\text{Revenue} = \frac{\$25{,}000 + \$45{,}000}{40\%}$$

$$= \frac{\$70{,}000}{0.4}$$

$$= \$175{,}000$$

Company goals would be set for $175,000 in new contracts.

Cost-Plus Pricing and Profit

Some managers would elect the absorption approach to cost-plus pricing and profit; others would use the contribution margin approach.

The absorption method would look as shown in the following example:

Direct materials	$ 8,000
Direct labor	6,000
Overhead at $3/DLH	1,500
Total cost to construct project	$15,500
Markup to cover selling and administrative expenses and desired profit—50% of cost to manufacture	$ 7,750
Target contract price	$23,250

The contribution method would look as shown in the following examples:

Direct materials	$ 8,000
Direct labor	6,000
Variable overhead	1,500
Variable selling and administrative expenses	750
Total variable expenses	$16,250

(*continued*)

Markup to cover fixed expenses and desired profit—43.1% of total variable expenses	$ 7,000
Target contract price	$23,250

Although the numbers vary, the actual profit is the same using either method. The markup in the absorption method takes into account that selling and administrative costs are considered a function of fixed costs, whereas in the example using the contribution margin approach, the accountant makes an effort to divide selling and administrative costs as *part* variable and assignable to this contract. The remainder of these costs are listed as fixed costs together with any other fixed costs.

Profit Objective and Cost-Plus Pricing

The concepts of target net profit and cost-plus pricing can both be employed to define a contract price.

EXAMPLE 1:

Desired target net profit (25.7% X $175,000)	$ 45,000
Selling and administrative expenses	25,000
Total	$ 70,000 (a)
Cost to construct various jobs	$105,000 (b)
Required markup (a ÷ b)	66.7%

EXAMPLE 2: Suppose that we wish to bid on a contract (no. 830225). Our estimator and salesman compute the lowest bid price to meet the company's goal as follows:

Cost to construct	$36,000
Markup at 66.7% X 36,000	24,012
Minimum bid price	$60,012
Estimated profit gain	$15,435
and applied to fixed expenses	8,577

Since $60,012 = 34.3% of gross expected revenue, $15,435 represents 34.3% of the profit goal of $45,000. Subtracting the profit from the markup provides $8577 applied to fixed costs of $25,000 (also = 34.3%).

These are not all the pricing methods that can be used to set the company's profit program. However, they do show that contractors do not need to rely on gut feelings. Science can be applied. Targeting profit is a good idea, but the dollar value selected should include gains needed to meet capital improvements, return on investments, and if incorporated, a return on stockholders' investment, to name just a few important considerations.

REFERENCES, SOURCES, AND PRESENT VALUE TABLES

1. Publishers of financial, managerial accounting, and economic books.
   a. Prentice-Hall, Inc., Englewood Cliffs, NJ 07632
   b. Reston Publishing Co., Inc., 11480 Sunset Hills Road, Reston, VA 22090
   c. McGraw-Hill Book Company, 1221 Avenue of the Americas, New York, NY 10020
   d. Business Publications, Inc., 200 Chisholm Place, Suite 240, Plano, TX 75075
   e. South-Western Publishing Company, 5101 Madison Road, Cincinnati, OH 45227
   f. American Institute of Certified Public Accountants, Inc. (AICPA), 1211 Avenue of the Americas, New York, NY 10036
   g. John Wiley & Sons, Inc., 605 Third Avenue, New York, NY 10158
   h. U.S. Department of Commerce, 14th Street between Constitution Avenue and East Street NW, Washington, DC 20230
   i. Robert Morris Associates, 1616 Philadelphia National Bank Building, Philadelphia, PA 19107

2. Educational sources
   a. Institutions
      (1) Business schools
      (2) Junior colleges
      (3) Senior colleges
      (4) Night school
      (5) College professors
      (6) CPAs
      (7) Financial institutions
   b. Subjects
      (1) Accounting general
      (2) Economics; and economics for the firm
      (3) Financial accounting
      (4) Managerial accounting
      (5) Banking and money
      (6) Real estate financing
      (7) Statistics
      (8) Financial forecasting
      (9) Accounting for the firm

(10) Advanced accounting fundamentals
(11) Accounting with computers

3. Sources of working tools
   a. Software programs
   b. Microcomputer equipment; hand-held calculators
   c. Accounting services

TABLE 2-1 PRESENT VALUE OF $1 RECEIVED AT END OF *N* PERIODS:

$$PVIF_{kt} = \frac{1}{(1+k)^t}$$

Periods until payment	1%	2%	4%	6%	8%	10%	12%	14%	15%	16%	18%
1	0.990	0.980	0.962	0.943	0.926	0.909	0.893	0.877	0.870	0.862	0.847
2	0.980	0.961	0.925	0.890	0.857	0.826	0.797	0.769	0.756	0.743	0.718
3	0.971	0.942	0.889	0.840	0.794	0.751	0.717	0.675	0.658	0.641	0.609
4	0.961	0.924	0.855	0.792	0.735	0.683	0.636	0.592	0.572	0.552	0.516
5	0.951	0.906	0.822	0.747	0.681	0.621	0.567	0.519	0.497	0.476	0.437
6	0.942	0.888	0.790	0.705	0.630	0.564	0.507	0.456	0.432	0.410	0.370
7	0.933	0.871	0.760	0.665	0.583	0.513	0.452	0.400	0.376	0.354	0.314
8	0.923	0.853	0.731	0.627	0.540	0.467	0.404	0.351	0.327	0.305	0.266
9	0.914	0.837	0.703	0.592	0.500	0.424	0.361	0.308	0.284	0.263	0.228
10	0.905	0.820	0.676	0.558	0.463	0.386	0.322	0.270	0.247	0.227	0.191
11	0.896	0.804	0.650	0.527	0.429	0.350	0.287	0.237	0.215	0.195	0.162
12	0.887	0.788	0.625	0.497	0.397	0.319	0.257	0.208	0.187	0.168	0.137
13	0.879	0.773	0.601	0.469	0.368	0.290	0.229	0.182	0.163	0.145	0.116
14	0.870	0.758	0.577	0.442	0.340	0.263	0.205	0.160	0.141	0.125	0.099
15	0.861	0.743	0.555	0.417	0.315	0.239	0.183	0.140	0.123	0.108	0.084
16	0.853	0.728	0.534	0.394	0.292	0.218	0.163	0.123	0.107	0.093	0.071
17	0.844	0.714	0.513	0.371	0.270	0.198	0.146	0.108	0.093	0.080	0.060
18	0.836	0.700	0.494	0.350	0.250	0.180	0.130	0.095	0.081	0.069	0.051
19	0.828	0.686	0.475	0.331	0.232	0.164	0.116	0.083	0.070	0.060	0.043
20	0.820	0.673	0.456	0.312	0.215	0.149	0.104	0.073	0.061	0.051	0.037
21	0.811	0.660	0.439	0.294	0.199	0.135	0.093	0.064	0.053	0.044	0.031
22	0.803	0.647	0.422	0.278	0.184	0.123	0.083	0.056	0.046	0.038	0.026
23	0.795	0.634	0.406	0.262	0.170	0.112	0.074	0.049	0.040	0.033	0.022
24	0.788	0.622	0.390	0.247	0.158	0.102	0.066	0.043	0.035	0.028	0.019
25	0.780	0.610	0.375	0.233	0.146	0.092	0.059	0.038	0.030	0.024	0.016
26	0.722	0.598	0.361	0.220	0.135	0.084	0.053	0.033	0.026	0.021	0.014
27	0.764	0.586	0.347	0.207	0.125	0.076	0.047	0.029	0.023	0.018	0.011
28	0.757	0.071	0.333	0.196	0.116	0.069	0.042	0.026	0.002	0.016	0.010
29	0.749	0.581	0.321	0.185	0.107	0.063	0.037	0.022	0.017	0.014	0.008
30	0.742	0.552	0.308	0.174	0.099	0.057	0.033	0.020	0.015	0.012	0.007
40	0.672	0.453	0.208	0.097	0.046	0.022	0.011	0.005	0.004	0.003	0.001
50	0.608	0.372	0.141	0.054	0.021	0.009	0.003	0.001	0.001	0.001	

d. Firm's producing forms for easy computations
e. Consulting services

4. Present value tables
   a. Table 2-1: Present value of $1 received at the end of n periods.
   b. Table 2-2: Present value of $1 received annually for N years.
   c. Table 2-3: Future value of $1 at the end of n periods.

Periods until pay-ment	20%	22%	24%	25%	26%	28%	30%	35%	40%	45%	50%
1	0.833	0.820	0.806	0.800	0.794	0.781	0.769	0.741	0.714	0.690	0.667
2	0.694	0.672	0.650	0.640	0.630	0.610	0.592	0.549	0.510	0.476	0.444
3	0.579	0.551	0.524	0.512	0.500	0.477	0.455	0.406	0.364	0.328	0.296
4	0.482	0.451	0.423	0.410	0.397	0.373	0.350	0.301	0.260	0.226	0.198
5	0.402	0.370	0.341	0.328	0.315	0.291	0.269	0.223	0.186	0.156	0.132
6	0.335	0.303	0.275	0.262	0.250	0.227	0.207	0.165	0.133	0.108	0.088
7	0.279	0.249	0.222	0.210	0.198	0.178	0.159	0.122	0.095	0.074	0.059
8	0.233	0.204	0.179	0.168	0.157	0.139	0.123	0.091	0.068	0.051	0.039
9	0.194	0.167	0.144	0.134	0.125	0.108	0.094	0.067	0.048	0.035	0.026
10	0.162	0.137	0.116	0.107	0.099	0.085	0.073	0.050	0.035	0.024	0.017
11	0.135	0.112	0.094	0.086	0.079	0.066	0.056	0.037	0.025	0.017	0.012
12	0.112	0.092	0.076	0.069	0.062	0.052	0.043	0.027	0.018	0.012	0.008
13	0.093	0.075	0.061	0.055	0.080	0.040	0.033	0.020	0.013	0.008	0.005
14	0.078	0.062	0.049	0.044	0.039	0.032	0.025	0.015	0.009	0.006	0.003
15	0.065	0.051	0.040	0.035	0.031	0.025	0.020	0.011	0.006	0.004	0.002
16	0.054	0.042	0.032	0.028	0.035	0.019	0.015	0.008	0.005	0.003	0.002
17	0.045	0.034	0.026	0.023	0.020	0.015	0.012	0.006	0.003	0.002	0.001
18	0.038	0.028	0.021	0.018	0.016	0.012	0.009	0.005	0.002	0.001	0.001
19	0.031	0.023	0.017	0.014	0.012	0.009	0.007	0.003	0.002	0.001	
20	0.026	0.019	0.014	0.012	0.010	0.007	0.005	0.002	0.001	0.001	
21	0.022	0.015	0.011	0.009	0.008	0.006	0.004	0.002	0.001		
22	0.018	0.013	0.009	0.007	0.006	0.004	0.003	0.001	0.001		
23	0.015	0.010	0.007	0.006	0.005	0.003	0.002	0.001			
24	0.013	0.008	0.006	0.005	0.004	0.003	0.002	0.001			
25	0.010	0.007	0.005	0.004	0.003	0.003	0.001	0.001			
26	0.009	0.006	0.004	0.003	0.002	0.002	0.001				
27	0.007	0.005	0.003	0.002	0.002	0.001	0.001				
28	0.006	0.004	0.002	0.002	0.002	0.001	0.001				
29	0.005	0.003	0.002	0.002	0.001	0.001	0.001				
30	0.004	0.003	0.002	0.002	0.001	0.001					
40	0.001										
50											

TABLE 2-2 PRESENT VALUE OF $1 RECEIVED ANNUALLY FOR *N* YEARS:

$$PVIF_a = \sum_{t=1}^{N} \frac{1}{(1+k)^t}$$

Years (*N*)	1%	2%	4%	6%	8%	10%	12%	14%	15%	16%	18%
1	0.990	0.980	0.962	0.943	0.926	0.909	0.893	0.877	0.870	0.862	0.847
2	1.970	1.942	1.886	1.833	1.783	1.736	1.680	2.647	1.626	1.605	1.566
3	2.941	2.884	2,775	2.573	2.577	2.487	2.402	2.322	2.283	2.246	2.174
4	3.902	3.808	3.630	3.465	3.312	3.170	3.037	2.914	2.855	2.798	2.690
5	4.853	4.713	4.452	4.212	3.993	3.791	3.605	3.433	3.352	3.274	3.127
6	5.795	5.601	5.242	6.917	4.623	4.355	4.111	3.889	3.784	3.685	3.498
7	6.728	6.472	6.002	6.582	5.206	4.868	4.564	4.288	4.160	4.039	3.812
8	7.652	7.328	6.733	6.210	5.747	5.335	4.968	4.639	4.487	4.344	4.078
9	8.566	8.162	7.435	6.802	6.247	5.759	5.328	4.946	4.722	4.607	4.303
10	9.471	8.683	8.111	7.360	6.710	6.145	5.660	5.216	5.019	4.833	4.494
11	10.368	9.787	8.360	7.887	7.139	6.495	5.937	5.453	5.234	5.029	4.656
12	11.255	10.575	9.365	8.384	7.536	6.814	6.194	5.600	5.421	5.197	4.793
13	12.134	11.343	9.986	8.853	7.904	7.103	6.424	5.842	5.583	5.342	4.910
14	13.004	12.106	10.563	9.295	8.244	7.367	6.628	6.002	5.724	5.468	5.008
15	13.865	12.849	11.118	9.712	8.559	7.506	6.811	6.142	5.847	5.575	5.092
16	14.708	13.878	11.652	10.106	8.851	7.824	6.974	6.705	5.984	5.669	5.162
17	15.562	14.764	12.166	10.477	9.122	8.022	7.120	6.073	6.057	5.749	5.222
18	16.398	14.898	12.659	10.828	9.372	8.201	7.250	6.467	6.128	5.818	5.273
19	17.226	15.578	13.134	11.158	9.604	8.365	7.366	6.550	6.198	5.877	5.316
20	18.046	16.351	13.590	11.470	9.818	8.514	7.469	6.623	6.259	5.929	5.353
21	18.857	17.011	14.029	11.764	10.017	8.649	7.562	6.687	6.312	5.973	5.384
22	19.660	17.658	14.451	12.042	10.201	8.772	7.645	6.743	6.359	6.011	5.410
23	20.456	18.292	14.857	12.303	10.371	8.883	7.718	6.792	6.399	6.044	5.432
24	21.243	18.914	15.247	12.550	10.529	8.985	7.784	6.835	6.434	6.073	5.451
25	22.023	19.523	15.622	12.783	10.675	9.077	7.843	6.873	6.464	6.097	5.467
26	22.795	20.121	15.983	13.003	10.810	9.161	7.896	6.906	6.491	6.118	5.480
27	23.560	20.807	16.330	13.211	10.935	9.287	7.943	6.935	6.514	6.136	5.492
28	24.316	21.881	16.663	13.406	11.051	9.307	7.984	6.961	6.534	6.152	5.502
29	25.066	21.844	16.984	13.591	11.158	9.870	8.022	6.983	6.551	6.166	5.510
30	25.818	12.395	17.292	13.765	11.258	9.227	8.055	7.003	6.566	6.177	5.517
40	32.835	17.353	19.793	15.046	11.925	9.779	8.244	7.105	6.642	6.234	5.548
50	39.196	31.435	21.182	15.762	12.234	6.915	8.604	7.133	6.661	6.246	5.554

Years (N)	20%	22%	24%	25%	26%	28%	30%	35%	40%	45%	50%
1	0.833	0.820	0.806	0.800	0.794	0.781	0.769	0.741	0.714	0.690	0.667
2	1.528	1.492	1.457	1.440	1.424	1.392	1.361	1.289	1.224	1.165	1.111
3	2.106	2.042	1.981	1.952	1.923	1.868	1.816	1.696	1.589	1.493	1.407
4	2.589	2.494	2.404	2.362	2.320	2.241	2.166	1.997	1.849	1.720	1.605
5	2.991	2.864	2.745	2.689	2.635	2.532	2.436	2.220	2.035	1.876	1.737
6	3.326	3.167	3.020	2.951	2.885	2.759	2.643	2.385	2.168	1.983	1.824
7	3.605	3.416	3.242	3.161	3.083	2.937	2.802	2.508	2.263	2.057	1.883
8	3.837	3.619	3.421	3.329	3.241	3.076	2.925	2.598	2.331	2.108	1.922
9	4.031	3.786	3.506	3.463	3.366	3.184	3.019	2.665	2.379	2.144	1.948
10	4.192	3.923	3.682	3.571	3.465	3.269	3.092	2.715	2.414	2.168	1.965
11	4.327	4.035	3.776	3.656	3.544	3.335	3.147	2.752	2.438	2.185	1.977
12	4.439	4.127	3.851	3.725	3.606	3.387	3.190	2.779	2.456	2.196	1.985
13	4.533	4.203	3.912	3.780	3.656	3.427	3.223	2.799	2.468	2.204	1.990
14	4.611	4.265	3.962	3.824	3.695	3.459	3.249	2.814	2.477	2.210	1.993
15	4.678	4.315	4.001	3.859	3.726	3.483	3.268	2.825	2.484	2.214	1.995
16	4.730	4.357	4.033	3.887	3.751	3.503	3.283	2.834	2.489	2.216	1.997
17	4.785	4.391	4.059	3.910	3.771	3.518	3.295	2.840	2.492	2.218	1.998
18	4.812	4.419	4.080	3.928	3.386	3.529	3.304	2.844	2.494	2.219	1.999
19	4.814	4.442	4.097	3.942	3.799	3.539	3.311	2.848	2.496	2.220	1.999
20	4.870	4.460	4.110	3.954	3.808	3.546	3.316	2.850	2.497	2.221	1.999
21	4.891	4.476	4.121	3.963	3.816	3.551	3.320	2.852	2.498	2.221	2.000
22	4.909	4.488	4.130	3.970	3.822	3.556	3.323	2.853	2.498	2.222	2.000
23	4.925	4.499	4.137	3.976	3.827	3.559	3.325	2.854	2.499	2.222	2.000
24	4.937	4.507	4.143	3.981	3.831	3.562	3.327	2.855	2.499	2.222	2.000
25	4.948	4.514	4.147	3.985	3.834	3.564	3.329	2.856	2.499	2.222	2.000
26	4.956	4.520	4.151	3.988	3.837	3.566	3.330	2.856	2.500	2.222	2.000
27	4.964	4.524	4.154	3.990	3.839	3.567	3.331	2.856	2.500	2.222	2.000
28	4.970	4.528	4.157	3.992	3.840	3.568	3.331	2.857	2.500	2.222	2.000
29	4.975	4.531	4.159	3.994	3.841	3.569	3.332	2.857	2.500	2.222	2.000
30	4.979	4.534	4.160	3.995	3.842	3.569	3.332	2.857	2.500	2.222	2.000
40	4.997	4.544	4.166	3.999	3.846	3.571	3.333	2.857	2.500	2.222	2.000
50	4.999	4.545	4.167	4.000	3.846	3.571	3.333	2.857	2.500	2.222	2.000

TABLE 2-3 FUTURE VALUE OF $1 AT THE END OF *N* PERIODS: $FVIF_{k,n} = (1 + k)^n$

Period	1%	2%	3%	4%	5%	6%	7%	8%	9%	10%
1	1.0100	1.0200	1.0300	1.0400	1.0500	1.0600	1.0700	1.0800	1.0900	1.1000
2	1.0201	1.0404	1.0609	1.0816	1.1025	1.1236	1.1449	1.1664	1.1881	1.2100
3	1.0303	1.0612	1.0927	1.1249	1.1576	1.1910	1.2250	1.2597	1.2950	1.3310
4	1.0406	1.0824	1.1255	1.1699	1.2155	1.2625	1.3108	1.3605	1.4116	1.4641
5	1.0510	1.1041	1.1593	1.2167	1.2763	1.3382	1.4026	1.4693	1.5386	1.6105
6	1.0615	1.1262	1.1941	1.2653	1.3401	1.4185	1.5007	1.5869	1.6771	1.7716
7	1.0721	1.1487	1.2299	1.3159	1.4071	1.5036	1.6058	1.7138	1.8280	1.9487
8	1.0829	1.1717	1.2668	1.3686	1.4775	1.5938	1.7182	1.8509	1.9926	2.1436
9	1.0937	1.1951	1.3048	1.4233	1.5513	1.6895	1.8385	1.9990	2.1719	2.3579
10	1.1046	1.2190	1.3439	1.4802	1.6289	1.7908	1.9672	2.1589	2.3674	2.5937
11	1.1157	1.2434	1.3842	1.5395	1.7103	1.8983	2.1049	2.3316	2.5804	2.8531
12	1.1268	1.2682	1.4258	1.6010	1.7959	2.0122	2.2522	2.5182	2.8127	3.1384
13	1.1381	1.2936	1.4685	1.6651	1.8856	2.1329	2.4098	2.7196	3.0658	3.4523
14	1.1495	1.3195	1.5126	1.7317	1.9799	2.2609	2.5785	2.9372	3.3417	3.7975
15	1.1610	1.3459	1.5580	1.8009	2.0789	2.3966	2.7590	3.1722	3.6425	4.1772
16	1.1726	1.3728	1.6047	1.8730	2.1829	2.5404	2.9522	3.4259	3.9730	4.4950
17	1.1843	1.4002	1.6528	1.9479	2.2920	2.6928	3.1588	3.7000	4.3276	5.0545
18	1.1961	1.4282	1.7024	2.0258	2.4066	2.8543	3.3799	3.9960	4.7171	5.5599
19	1.2081	1.4568	1.7535	2.1068	2.5270	3.0256	3.6165	4.3157	5.1417	6.1159
20	1.2202	1.4859	1.8061	2.1911	2.6533	3.2071	3.8697	4.6610	5.6044	6.7275
21	1.2324	1.5157	1.8603	2.2788	2.7860	3.3996	4.1406	5.0338	6.1088	7.4002
22	1.2447	1.5460	1.9161	2.3699	2.9253	3.6035	4.4304	5.4365	6.6586	8.1403
23	1.2572	1.5769	1.9736	2.4647	3.0715	3.8197	4.7405	5.8715	7.2579	8.9543
24	1.2697	1.6084	2.0328	2.5633	3.2251	4.0489	5.0724	6.3412	7.9111	9.8497
25	1.2824	1.6406	2.0938	2.6658	3.3864	4.2919	5.4274	6.8485	8.6231	10.834
26	1.2953	1.6734	2.1566	2.7725	3.5557	4.5494	5.8074	7.3964	9.3992	11.918
27	1.3082	1.7069	2.2213	2.8834	3.7335	4.8223	6.2139	7.9881	10.245	13.110
28	1.3213	1.7410	2.2879	2.9987	3.9201	5.1117	6.6488	8.6271	11.167	14.421
29	1.3345	1.7758	2.3566	3.1187	4.1161	5.4184	7.1143	9.3173	12.172	15.863
30	1.3478	1.8114	2.4273	3.2434	4.3219	5.7435	7.6123	10.062	13.267	17.449
40	1.4889	2.2080	3.2620	4.8010	7.0400	10.285	14.974	21.724	31.409	45.259
50	1.6446	2.6916	4.3839	7.1067	11.467	18.420	29.457	46.901	74.357	117.39
60	1.8167	3.2810	5.8916	10.519	18.679	32.987	57.946	101.25	176.03	304.48

*FVIF > 99,999

Period	12%	14%	15%	16%	18%	20%	24%	28%	32%	36%
1	1.1200	1.1400	1.1500	1.1600	1.1800	1.2000	1.2400	1.2800	1.3200	1.3600
2	1.2544	1.2996	1.3225	1.3456	1.3924	1.4400	1.5376	1.6384	1.7424	1.8496
3	1.4049	1.4815	1.5209	1.5609	1.6430	1.7280	1.9066	2.0972	2.3000	2.5155
4	1.5735	1.6890	1.7490	1.8106	1.9388	2.0736	2.3642	2.6844	3.0360	3.4210
5	1.7623	1.9254	2.0114	2.1003	2.2878	2.4883	2.9316	3.4360	4.0075	4.6526
6	1.9738	2.1950	2.3131	2.4364	2.6996	2.9860	3.6352	4.3980	5.2899	6.3275
7	2.2107	2.5023	2.6600	2.8262	3.1855	3.5832	4.5077	5.6295	6.9826	8.6054
8	2.4760	2.8526	3.0590	3.2784	3.7589	4.2998	5.5895	7.2058	9.2170	11.703
9	2.7731	3.2519	3.5179	3.8030	4.4355	5.1598	6.9310	9.2234	12.166	15.916
10	3.1058	3.7072	4.0456	4.4114	5.2338	6.1917	8.5944	11.805	16.059	21.646
11	3.4785	4.2262	4.6524	5.1173	6.1759	7.4301	10.657	15.111	21.198	29.439
12	3.8960	4.8179	5.3502	5.9360	7.2876	8.9161	13.214	19.342	27.982	40.037
13	4.3635	5.4924	6.1528	6.8858	8.5994	10.699	16.386	24.758	36.937	54.451
14	4.8871	6.2613	7.0757	7.9875	10.147	12.839	20.319	31.691	48.756	74.053
15	5.4736	7.1379	8.1371	9.2655	11.973	15.407	25.195	40.564	64.348	100.71
16	6.1304	8.1372	9.3576	10.748	14.129	18.488	31.242	51.923	84.953	136.96
17	6.8660	9.2765	10.761	12.467	16.672	21.186	38.740	66.461	112.13	186.27
18	7.6900	10.575	12.375	14.462	19.673	26.623	48.038	85.070	148.02	253.33
19	8.6128	12.055	14.231	16.776	23.214	31.948	59.567	108.89	195.39	344.53
20	9.6463	13.743	16.366	19.460	27.393	38.337	73.864	139.37	257.91	468.57
21	10.803	15.667	18.821	22.574	32.323	46.005	91.591	178.40	340.44	637.26
22	12.100	17.861	21.644	26.186	38.142	55.206	113.57	228.35	449.39	866.67
23	13.552	20.361	24.891	30.376	45.007	66.247	140.83	292.30	593.19	1178.6
24	15.178	23.212	28.625	35.236	53.108	79.496	174.63	374.14	783.02	1602.9
25	17.000	26.461	32.918	40.874	62.668	95.396	216.54	578.90	1033.5	2180.0
26	19.040	30.166	37.856	47.414	73.948	114.47	268.51	612.99	1364.3	2964.9
27	21.324	34.389	43.535	55.000	87.259	137.37	332.95	784.63	1800.9	4032.2
28	23.883	39.204	50.065	63.800	102.96	164.84	412.86	1004.3	2377.2	5483.8
29	26.749	44.693	57.575	74.008	121.50	197.81	511.95	1285.5	3137.9	7458.0
30	29.959	50.950	66.211	85.849	143.37	237.37	634.81	1645.5	4142.0	10143.
40	93.050	188.88	267.86	378.72	750.37	1469.7	5455.9	19426.	66520.	*
50	289.00	700.23	1083.6	1670.7	3927.3	9100.4	46890.	*	*	*
60	897.59	2595.9	4383.9	7370.1	20555.	56347.	*	*	*	*

Robert Morris Associates Disclaimer: Interpretation of Statement Studies Figures

RMA recommends that Statement Studies data be regarded only as general guidelines and not as absolute industry norms. There are several reasons why the data may not be fully representative of a given industry:

1. The financial statements used in the *Statement Studies* are not selected by any random or statistically reliable method. RMA member banks voluntarily submit the raw data they have available each year, with these being the only constraints: (a) The fiscal year-ends of the companies reported may not be from April 1 through June 29, and (b) their total assets must be less than $100 million.
2. Many companies have varied product lines; however, the *Statement Studies* categorize them by their primary product Standard Industrial Classification (SIC) number only.
3. Some of our industry samples are rather small in relation to the total number of firms in a given industry. A relatively small sample can increase the chances that some of our composites do not fully represent an industry.
4. There is the chance that an extreme statement can be present in a sample, causing a disproportionate influence on the industry composite. This is particularly true in a relatively small sample.
5. Companies within the same industry may differ in their method of operations which in turn can directly influence their financial statements. Since they are included in our sample, too, these statements can significantly affect our composite calculations.
6. Other considerations that can result in variations among different companies engaged in the same general line of business are different labor markets; geographical location; different accounting methods; quality of products handled; sources and methods of financing; and terms of sale.

For these reasons, RMA does not recommend the *Statement Studies* figures be considered as absolute norms for a given industry. Rather the figures should be used only as general guidelines and in addition to the other methods of financial analysis. RMA makes no claim as to the representativeness of the figures printed in this book.

3

ACCOUNTING AND BOOKKEEPING

QUICK REFERENCES

In this chapter many of the forms and formats used by bookkeepers and accountants are identified. As a lead-in to the chapter, information is provided on the basics of bookkeeping and accounting. For the beginning contractor this information may be the first inkling of the complex system or systems required to keep a sound and visual understanding of how the enterprise (company or firm) is doing. For example, can you now answer these questions with any accuracy?

1. Are you making a profit?
2. Is it as high as you would like?
3. Are you meeting your goals or the company's goals?

4. Where, if at all, is there uncontrolled waste and inefficiency?
5. Is the company in sound financial condition or just hanging on?

These are questions that can be answered if good accounting principles are used; and it takes good bookkeeping to provide the inputs. Therefore, *bookkeeping* is the process of making entries for all incoming and outgoing funds. This means that ledgers and journals must be established and used; transactions must be catalogued, and bills must be sent out and others paid. *Accounting*, on the other hand, is concerned with the results of the hundreds of bookkeeping entries. In accounting, all transactions are usually accounted for by the simplified or double-entry systems. The reports generated by these entries are extremely meaningful to sole proprietors, partners, corporate managers, and stockholders. These reports also provide the basis for computing the financial ratios illustrated in Chapter 2.

CASH, ACCRUAL, AND CONSTRUCTION CONTRACT ACCOUNTING METHODS

Cash Method

When the cash method of accounting is used in an enterprise, *income* is accounted for in the year in which it is received as cash or property. This means, for example, that a deposit received from a contract is income during that accounting year. The deposit is cash since it is spendable. Property obtained by the enterprise for services rendered, for example, is considered income at its *fair market value*.

Expenses under the cash method are deductible in the year in which they are paid. For example, the payment of a supply house invoice of $2000 represents a cash expense. Suppose that you borrow the $2000 to pay the invoice; it is still considered a cash expense for the current year. However, the repayment of the loan in later years will not be an expense.

Interestingly, a contractor issuing a check considers it an expense on the date of issue, whereas the recipient does not consider it income on the issue date but rather on the date the bank accepts the check (assuming that it is a good check).

Also, cash paid out for capital goods may be deducted over a period of years through annual depreciation charges.

Accrual Method

When the accrual method of accounting is used in an enterprise, *income* is accounted for when the right to receive it comes into being. There must be an unconditional right to receive the income. For example, a carpenter installs a storm door on Mr. Jones' house. After completing the job to Mr. Jones' satisfaction, the carpenter presents a bill and lets Mr. Jones have 30 days to pay. The carpenter has income accrued when he submits his bill, not 30 days later when he actually receives Mr. Jones' check.

Similarly, an expense occurs when all the events have occurred within the taxable year surrounding the expense. This is often termed *when the liability has occurred.* Let's explain. Suppose that a contractor orders a variety of building supplies and receives them piecemeal. He accrues the expense after the last delivery. In this situation the contractor's contract with the building-supply house is not complete until the last delivery is made—the events of the liability must be complete.

The only problems arising from the accrual method are those at the end of the accounting year and extending into the next year. Consult your tax accountant for explanations or read various accounting literature on this subject (see the References and Sources at the end of the chapter).

Completed-Contract and Percentage-of-Completion Methods

When employing the *completed-contract* method of accounting, the net profit on the entire job may be reported in the year when the job is completed and accepted, even if it was started in one accounting year and completed in the next. This means that all incomes and expenses are deferred, as are contract-related overhead and depreciation expenses.

When adopting the *percentage-of-completion* method, only the part of the profit earned as a percentage of contract completion is recorded as income for the year. Let us say that a contract for erecting an apartment is $1.5 million and is expected to take 15 months to complete. Construction begins in June and is 60% complete (verifiable) at year's end. Profit is calculated as 60% of expected profit. The next year the contract is completed and the remaining 40% profit is computed. Under this method, overhead and expenses, depreciation, and so on, are allocated by the same percentages.

Special Note. Both of these methods have special limitations under U.S. tax law regulations.

1. The contract period extends into more than one accounting year.
2. Stockpiling of materials is expected.
3. The contract may have a clause withholding a percentage of payments until contract completion.
4. The need to order long-time special building items (prepaying or C.O.D.) distorts profits by increasing expenses in one year or the other.

ACCOUNTING TERMS AND DEPRECIATION METHODS

Terms

Assets, Liabilities, and Capital. *Assets* are the properties owned by the business enterprise (owners). They may include *cash, accounts receivable, inventories or supplies, prepaid accounts, land, equipment, and real property.*

Liabilities are debts owed by the business enterprise to others and may include accounts payable (to building-supply houses, etc.), *notes payable (loans), taxes payable, salaries payable, and others.*

Capital, also referred to as *owners' equity,* is that part of the enterprise's assets not obligated to liabilities. Thus we see that

$$\text{Assets} = \text{liabilities} + \text{capital} \tag{3-1}$$

Expenses. Frequently termed *operating expenses,* these are transactions that account for funds being spent for services rendered to the business, depreciation, and taxes to be paid out for employees, among other things.

Income or Revenue. *Income* or *revenue* is the term used any time that money or property is received by the business. It may be cash or checks, as the fair market value of property. It may also be called income from operations, or income from joint ventures.

Net Income. *Net income* is the remainder of revenue less expenses. Where revenue is greater than expenses, net income is a profit; where the opposite is true, net income represents a loss. Net income is usually a value subject to tax.

Transactions. A *transaction* is an occurrence or event of a condition that must be recorded. For example, they may include the payment of the monthly rent, utilities, or an accounts payable such as a lumber bill. A transaction flows into and out of the enterprise as follows:

Transaction ---> Business document ---> Journal Entry ---> Ledger Entry Posted ---> Bill paid or Invoice submitted for payment

Retained Earnings. The net income after tax that the enterprise earns which either (1) adds to capital, (2) may be used to increase real or other property, (3) may be used to pay dividends, or (4) may be used to purchase treasury stock is called *retained earnings.*

Withdrawals. All sole proprietorships and unincorporated partnerships use *drawing accounts. Withdrawals* by the sole proprietor (or partners) are salaries paid to themselves and are taken from net income. When the withdrawal is made, the amount is entered into the drawing account.

Depreciation Methods

The depreciation is the accounting for, over a predetermined life of the asset, the cost of the asset. It shows the decrease in value the asset has to the enterprise, and defers taxes associated with the asset to later periods. The total sums of depreciation over a period of years associated with an asset (equipment) is referred to as *accumulated depreciation.* Several forms of depreciation are: straight-line, 150% declining balance, 175% declining balance, double declining balance, sum of the years' digits, and accelerated cost recovery system. Examples for computing depreciation follow the definitions and formulas.

Straight-Line Method. The formula for the straight-line method is

$$\text{Depreciation/year} = \frac{\text{cost (less salvage)}}{N} \tag{3-2}$$

$$\text{Depreciation rate} = \frac{1}{N} \tag{3-3}$$

Where N represents the number of years the asset will be depreciated (expected life).

EXAMPLE: A piece of construction equipment cost $1200, is expected to have no salvage value, and will last 5 years.

$$\text{Depreciation/year} = \frac{\$1200}{5}$$

$$= \$240 \tag{3-2}$$

or

$$\text{Depreciation/year} = \frac{1}{5}$$

$$= 0.2 \text{ or } 20\% \text{ per year} \tag{3-3}$$

Therefore,

$$0.2 \times \$1200 = \$240 \text{ depreciation/year}$$

Declining Balance Methods. Formulas for the declining balance methods are

1. Double declining balance:

$$\text{Depreciation} = \frac{\text{book value}}{N} \times 200\% \tag{3-4}$$

$$\text{Depreciation rate} = \frac{1}{N} \times 200\% \tag{3-5}$$

Where N represents the years the asset will be depreciated.

2. 175% declining balance:

$$\text{Depreciation} = \frac{\text{book value}}{N} \times 175\% \tag{3-6}$$

$$\text{Depreciation rate} = \frac{1}{N} \times 175\% \tag{3-7}$$

3. 150% declining balance:

$$\text{Depreciation} = \frac{\text{book value}}{N} \times 150\% \tag{3-8}$$

$$\text{Depreciation rate} = \frac{1}{N} \times 150\% \tag{3-9}$$

Note: These declining balance methods do *not* develop an accumulated depreciation equal to the asset's original book value at the end of the depreciating life. The residue is available to apply to salvage, it may be added to the final year as "out," or it may be carried to the next year.

EXAMPLE: Let us use as a sample problem a $1200 piece of equipment with a 5-year life.

Double declining balance (DDB)

$$\text{Depreciating rate/year} = \frac{1}{N} \times 200\% \tag{3-4}$$

Year	Book value	DDB rate	Depr.	Accum depr.	End-of-year book value
1	$1200	40%	$480	—	$720.
2	$ 720	40%	$288	$ 480	$432
3	$ 432	40%	$172.00	$ 786	$259.20
4	$ 259.20	40%	$103.68	$ 840.80	$155.52
5	$ 155.52	40%	$ 62.08	$1044.48	$ 93.44

Sum-of-the-Years'-Digits Method. The formula for the sum-of-the-years'-digits method is

$$\text{Depreciation} = \text{cost (less salvage)} \times \frac{N_i}{\text{sum of the years' digits}} \tag{3-10}$$

where N_i represents the number of estimated life years remaining.

$$\text{Depreciation rate} = \frac{N_i}{N\left(\frac{N+1}{2}\right)} \tag{3-11}$$

where N = number of years of estimated life

N_i = number of estimated life years remaining

EXAMPLE:

Year	Cost less salvage	Rate	Depreciation	Book value
1	$1200	5/15	$400	$800
2	1200	4/15	320	480
3	1200	3/15	240	240
4	1200	2/15	160	80
5	1200	1/15	80	0

Accelerated Cost Recovery System (ACRS). Most property placed into service after 1980 is recovery property (tangible property of a character subject to depreciation). Table 3-1 illustrates the accelerated cost recovery system.

TABLE 3-1 ACCELERATED COST RECOVERY SYSTEM

Recovery period	Type of property	Depreciation	Percentage
3-year property	Automobile, light trucks	1st yr.	25%
	Special manufacturing	2nd yr.	38%
	Tools	3rd yr.	37%
5-year property	Equipment, office furniture and fixtures	1st yr.	15%
		2nd yr.	22%
		3rd–5th yr.	21%
10-year property	Certain real and utility property, manufactured homes, mobile homes	1st yr.	8%
		2nd yr.	14%
		3rd yr.	12%
		4th–6th yr.	10%
		7th–10th yr.	9%
15-year property	Real property	This depends on the month of the year property was placed in service; IRS has compiled a table of allowable depreciation, e.g.:	

Month:	7	8
1st yr.	6%	5%
2nd yr.	11%	11%
3rd yr.	10%	10%
4th yr.	9%	9%

EXAMPLE: In January a company buys a cement mixer for $1200 in 19X3 and elects to use ACRS 3-year property. The unadjusted basis is $1200.

Class of property	Date placed in service	Cost	Recovery period	Method of figuring depr.	Percentage	Depr. this year
3-year	19X3	$1200	3	ACRS	25	$ 300
	19X4	1200	3	ACRS	38	456
	19X5	1200	3	ACRS	37	444
					100%	$1200

Note: There are alternatives to the ACRS method described above. Refer to IRS Publications 334 and 534.

JOURNAL, LEDGER, AND REGISTER FORMS

Various books and records are used to keep track of all transactions of a company. We shall examine some of the various forms of *journals*, *ledgers*, and *registers*.

Journals

A book that contains two, three, four, or more columns used to record a transaction is called a journal. Several forms are cash receipts, check disbursement, general, and purchase journal. There are several other related journal entries that are used in ledgers.

Cash Receipts Journal Form. An example of a five-column form is shown below. Notice that cash received from any source is listed in this journal and the cash column shows debits, while other columns are used to show that creditors paid their debts.

CASH RECEIPTS JOURNAL Page______

Date	Account Credited	Post. Ref.	Sundry Accounts (CR)	Revenue (CR)	Contracts Receivable	Discounts (DR)	Cash (DR)
	Notes Rec.						
	Interest Inc.						
	A.B. Const. Co.						
	Purchases						
	Contract						
	C3–82						

Check Disbursement Journal Form. This is a journal that has many columns which are designed to make easy work of recording check disbursement. Of course, the heading of each column would be across the top of the journal; however, we show the form partially developed and list all column headings separately as follows:

CHECK DISBURSEMENT JOURNAL Page ______

Year Date	Paid To	Ck No.	Amount of Ck.	Materials	General Accounts	
					Title	Amount

Column headings between Materials and General Accounts may include:

Materials	Electric
Gross payroll	Interest
FITW	Rent
FICA reserve	Telephone
State-withheld income tax	Truck/auto
Employer's FICA tax	Drawing
	General accounts

Cash Payments Journal Form. The more traditional form of check disbursement is called the cash payments journal (form). An example of a four-column form is shown below.

CASH PAYMENTS JOURNAL Page ______

Date	Ck. No.	Account Debited	Post. Ref.	Sundry (DR)	Accounts Payable (DR)	Purchased Discounts (CR)	Cash (CR)

General Journal Form. The general journal is customarily a two-column form. The general journal is used for miscellaneous entries that do not fit the special journals. An example of a two-column form is shown below.

GENERAL JOURNAL Page ______

Date	Description	Post. Ref.	Debit	Credit

Related Journal Entries That Are Used in Ledgers

1. *Adjusting entries* are those that are required at the end of the accounting period to record internal transactions (corrections to the ledgers). Accounts needing adjusting entries include:
   a. Inventory or supplies
   b. Prepaid rent and insurance
   c. Depreciation
   d. Salaries
2. *Closing entries* are used to "zero" or clear all the capital accounts. Closing entries are made in the ledgers and journal immediately after adjusting entries.
3. *Income summary*; an income summary is prepared in the *general journal* at the accounting period as part of the closing entries. This account summarizes the data in the revenue and expense accounts. Four entries are required for a sole proprietorship.
   a. The *revenue account (contract revenue) is debited.* The *income summary is credited.*
   b. The *income summary is debited* with the *sum of the totals of the expense accounts. Each expense account is credited* for the amount of its balance.
   c. The *income summary is debited* for the amount of its balance (net income). The *capital account is credited* with the same amount (net income). (*Note:* The entries are reversed if there is a net loss.)
   d. The *capital account is debited* the amount of the drawing account. The *drawing account is credited* for the amount in its balance.

   Note: Ledger the accounts after adjusting and closing entries have been journalized. Then prepare trial balances and accounting reports.
4. *Reversing entries* are needed where, for example, salaries of employees are accrued but not paid until sometime in the following fiscal period. A *reversing entry* is the opposite of an adjusting entry. These entries are made in the general journal first, then posted to the ledger accounts.

Purchase Journal Form. Three uses are made of the purchase journal. They include recording (1) materials for resale to contract customers (inventory items), (2) supplies for conducting the contracting company, and (3) equipment and plant assets. An example of a multicolumn purchase journal form is shown below.

PURCHASE JOURNAL Page ______

Date	Account Credited	Post. Ref.	Accounts Payable (CR)	Materials (DR)	Office Supplies (DR)	Job-site Supplies (DR)	Sundry Accounts		
							Account	Post Ref.	Amount
							Equipment Plant Asset		

Ledgers

A ledger is a book of related accounts that are considered to be a complete unit. A ledger may contain controlling accounts or subsidiary accounts. The *general ledger* usually contains the controlling accounts. Individual parties, customers, and related entries of a common nature are input to one of the various subsidiary ledgers. Each subsidiary ledger provides the input to its controlling account in the general ledger.

General Ledger Accounts and Forms. Some general ledger accounts include:

Assets

100 Cash
105 Contracts receivable
130 Supplies
135 Materials (inventory)
140 Prepaid items
160 Equipment
160.1 Accumulated depreciation

Liabilities

200 Accounts payable
205 Contracts payable
220 Employee payroll deductions
225 Salaries payable
230 Taxes payable

Capital

800 Stock (common)
810 Stock (preferred)
820 Drawing
850 Capital
850 Retained earnings
890 Income summary

Indirect expenses

600 Supervision labor
610 Indirect labor
625 Tools
650 Office expenses
660 Professional fees
665 Rent expense

Revenue

300 Revenue

Long-term debt

700 Notes payable

Operating expenses

500 Contract expenses
530 Job-site expenses
590 Miscellaneous expense

An example of a 4-column form is shown below.

Account Contracts Receivable Acc. No. 113

Date	Item	Post. Ref.	Debit	Credit	Balance	
					Debit	Credit

Subsidiary Ledgers and Forms. Several subsidiary ledgers would be:

1. All contractors that provide subcontracting services
2. All creditors
3. Various vendors who supply materials
4. Stockholder of record
5. Subscribers of record

An Example of a three-column ledger is shown below.

Name

Address					
Date	Item	Post. Ref.	Debit	Credit	Balance

Let us illustrate an example of subsidiary accounts and controlling account.

Subsidiary ledger entries contracts receivable		General ledger contracts receivable
		$4300
J. Ascot Construction	$ 300	
B. Glenn Construction Inc.	1200	
G. Harris Electrical Cont.	2200	
T. Kay Plumbing Cont.	600	
	$4300	

Registers

Registers are accounting forms that satisfy a need to record those entries that are repetitive in nature. Each company owner/manager needs to evaluate the type and variety of entries needed to give good visibility and ease of recording. Several example register forms are shown below.

Check Register Form. This register is used to show to whom every check written as payment is cataloged. The form looks like this:

CHECK REGISTER

No.	Date	Payee No.	Vou. No.	Accounts Payable (DR)	Purchase Dis. (CR)	Cash (CR)	Bank Deposits	Balance

Notes Register Form. Either a notes receivable or payable or both registers are used to record all data relating to notes. From these registers the notes receivable and/or payable accounts in the general ledger are posted. An example of a notes register is shown below.

NOTES ( )

Name of Maker/(Payee)	Place of Payment	Amount	Term	Interest Rate	Due Date

Payroll Register Form. A very important register and one easily handled by computer is the payroll register, shown below.

Example:

INSURANCE REGISTER

Date of policy	Policy No.	Insurer	Property or equipment	Amount	Term	Expiry date	Unexpired premium	Expired premium							Total	Unexpired premium
								Jan	Feb	Mar	Apr	May	Jun	. . . Dec		

Insurance Register Form. This register is very detailed since it shows all policies as well as monthly, quarterly, and/or yearly premium payments. From this register the "prepaid insurance" data can be calculated for use in the balance sheet. An example of an insurance register form is shown below.

Example:

PAYROLL REGISTER (WEEK ENDING – – – – – –)

Name of Employee	Tot Hrs	Earnings			Taxable Earnings		Deductions						Paid		Discount Debited	
		Reg	O.T.	Tot	Unemp. Comp.	FICA	FICA	FITW	FUTC	U.S. Savings Bonds	Misc.	Tot	Net Amt	Ck. No.	Workman's Pay Expense	Office Salaries

Voucher Register Form. Some contractors require that all transactions be accounted for through voucher systems. For example, a truck needs lubrication. The office staff issues a voucher to the truck driver to have the work done. The voucher is logged in a voucher register and all activities associated with the voucher are added as they occur. An example is shown below.

Example:

VOUCHER REGISTER

Date	Vou. No.	Payee	Paid		Accts. Payable	Purchases (DR)	Building Materials Exp.	Office Supply Exp.	Equip. Oper. Exp.	Misc. Office Exp.	Misc. Gen. Exp.	Sundry Accts		
			Date	Ck. No.								Account	Posting Ref.	Amount

ACCOUNTING STATEMENT FORMS

All ledgers and journals are convenient repositories of daily, weekly, and monthly transactions. However, they do not provide the sole proprietor, partner, or corporate manager or stockholder with information that is meaningful in terms of how well or poorly the enterprise is doing. Therefore, a variety of accounting statements are developed by using the variety of data available from the accounts.

Sole Proprietorship (Unincorporated) Forms

Three basic statements are developed at the end of the accounting period or at closer intervals if necessary:

1. The *income statement*
2. The *capital statement*
3. The *balance sheet*

Income Statement Form. The income statement is detailed to show all income, all expenses, and the difference between income and expense, called net income. An example is shown below.

ABC CONTRACTING[a] INCOME STATEMENT
FOR YEAR ENDED DECEMBER 31, 19X3

Contract revenues	
(itemized by job or consolidated)	$
Total revenues	
Less contract expenses	
(itemized by job or consolidated)	
Job-site indirect expenses	
Total contract expenses	
Gross profit on operations	$
General and administrative expenses	
Income from operations	
Other income and expenses	
Interest income on money market CDs	
Gain on sale of equipment	
Income before taxes (net income)[b]	$

[a]The name of the company is fictitious.

[b]Net income before taxes are paid.

Note: The entries are representative only and are not all the entries that may be included in the income statement.

Capital Statement Form. The *capital statement* illustrates the change in capital and how much the owner withdrew from the company for his or her salary. An example is shown below.

ABC CONTRACTOR CAPITAL STATEMENT
FOR YEAR ENDED DECEMBER 31, 19X3

Capital, January 1, 19X3	$
Net income for the year	
Less withdrawals	
Increase (decrease) in capital	
Capital, December 31, 19X3	$

Notes: 1. Net income is taken from the income statement.
2. Withdrawals are taken from the drawing account.
3. Capital January 1, 19X3, is taken from the capital account.

Balance Sheet Form. The third form the sole proprietor makes out is the *balance sheet.* This is a summary of assets, liabilities, and capital at the end of the reporting period. An example is shown below.

ABC CONTRACTOR BALANCE SHEET DECEMBER 31, 19X3

Assets	
Current assets	
Cash	
Contracts receivable	
Inventory	
Supplies	
Prepaid rent	
Prepaid insurance	
Total current assets	
Office and warehouse assets	
Office furniture and equipment	
Less accumulated depreciation	
Construction equipment and building	
Less accumulated depreciation	
Land	
Total assets	
Liabilities and capital	
Current liabilities	
Accounts payable	
Salaries payable	
Total current liabilities	
Notes payable	
Total liabilities	
ABC contractor, capital	
Total liabilities and capital	

Note: Total assets and total liabilities plus capital must balance.

Other Forms. A final note: There are many more forms available than those shown. It may be useful to use them and it may not. Each contractor should make a personal evaluation.

Some of them include:

1. Job status reports.
2. Job cost accounting
3. Individual accounts receivable and accounts payable
4. Notes payable report
5. Accounts receivable and accounts payable change or summary reports
6. Comparative income summaries, capital statements, and balance sheets
7. Funds-flow report showing changes in fund uses
8. Employee accounts
9. Tax accounts (FICA, social security), FITW (workman's compensation), FUTW (unemployment tax)
10. Income tax reports
11. Profit-and-loss summaries on each job

Partnership Forms

Partnerships may be unincorporated or incorporated enterprises. If unincorporated, they adopt most of the accounting principles of sole proprietorship. Each partner has a drawing account and at year's end they divide the net profit or income according to the Articles of Partnership division of income or percentage of capital investment in the enterprise. (Excess net income may be plowed back into the company as capital or for the purchase of assets.) If incorporated, the partners may, according to their position (director, business manager, silent partner, stockholder), obtain salaries and/or a percentage of net income or retained earnings based on the disposition of these funds. Several areas peculiar to partnerships are discussed in detail below.

Recording of Investments Forms. The *recording of investments* by the partners is done on separate entries in general journals. Various assets contributed by a partner are debited to those asset accounts. These may include cash, equipment, real estate, land, and so on. Similarly, liabilities, if any, are credited to the partner's accounts and, of course, the partner's capital account is credited with funds contributed. The partner's accounts—asset = liabilities + capital—must balance. The key here is that because partners often contribute different amounts and different types of assets and may bring in different liabilities, separate entries are needed for each partner.

Division of Income Forms. *Division of income* (net or loss) is a function of the original Articles of Partnership and may be handled several ways.

If partners are allocated specified salaries (those agreed to, based on their contribution of work to the enterprise), those amounts may be treated as either *drawing accounts* or *salary expenses.* Remaining net income is then distributed on an equal-share basis or on an agreed-upon basis. A percentage basis is usually employed when the partners' original or total contribution differ in dollar value.

EXAMPLE:

Net income $63,500.00

	Division of income to partner:		
	A	B	Total
Original investment	(60%)	(40%)	(100%)
Salary allowance	$24,000	$20,000	$44,000
Remaining income	11,700	7,800	19,500
Net income	$35,700	$27,800	$63,500

Return on Investment Forms. Some partnership agreements allow for a *return on investment* of their capital investment. If this is applicable, salaries are usually paid first, then the *remaining income* may be allocated to interest on investment, then any residue of remaining income is allocated on a percentage basis.

EXAMPLE:

Net income $65,500

	Division of income to partner:		
	A	B	Total
Original investment	(60%)	(40%)	(100%)
Salary expense	$24,000	$20,000	$44,000
Return on investment	1,386	924	2,310
Remaining income	11,514	7,676	19,190
Net income	$36,900	$28,600	$65,500

Partner A: 7% on an investment of $19,800 = $1386.

Partner B: 7% on an investment of $13,800 = $924.

In both examples the net income earned by each partner would be credited to their respective capital accounts. *Capital statements* would show the changes in capital for each partner.

Admission of an Additional Partner Forms. *Admission of an additional partner* dissolves the original agreements and creates new agreements. A new partner may be entered in either of two ways.

1. Through the purchase of an interest from one or more current partners. *Note:* The total capital of the company *does not* change. However, percentages of investment are adjusted. This means that *capital accounts* must be adjusted and a new *capital account* needs to be established for the new partner.
2. The new partner may bring new assets to the company. This increases capital in the company. A new account is begun for the new partner, adjustments to net income are specified for distribution, and various *asset accounts* are depicted to record the newly acquired assets. The *capital account* will be increased.

Withdrawal of a Partner Forms. *Withdrawal of a partner* is usually handled by the remaining partner or partners purchasing the withdrawing partners interests. The *capital accounting* adjustments require a debit to the withdrawing partners' capital account and a credit to the remaining partner(s)' capital accounts. A withdrawing partner usually requires a settlement of accounts. *Assets* are adjusted to current market value; then the percentage allocated to the withdrawing partner is paid for these assets. If assets are insufficient, the liability accounts are credited to indicate a debt remaining to be paid to the withdrawing partner.

Liquidation of the Partnership Forms. A *liquidation of the partnership* requires a full accounting of all accounts, and payments of all liabilities through conversion of assets to cash or other marketable assets. Residue, remaining income, if any, is apportioned according to distribution ratios. The sale of assets is called *realization.*

1. If realization results in gains (excess over book values of assets), the gains are apportioned according to distribution ratios.
2. If *realizations* result in losses, these are similarly apportioned according to distribution ratios. This loss must be made up by reducing the capital accounts of each partner. If the realizations (losses) cause the capital accounts to have *negative* balances, each partner must assume a personal debt according to his or her distribution ratio.

Death Effect on Accounting Forms. The *death* of a partner dissolves the partnership. The *accounts* are *closed* as of the date of death (unless provisions are made in the Articles of Partnership). *Net income* is transferred to the capital accounts. Since the business may continue to receive revenue, the accounts, including the deceased's, are kept open to allow for the accumulated earned income. The income and capital belonging to the deceased partner are transferred to the deceased's estate's *liability account.* Handling of the books will be treated either as they would for withdrawing partner or for liquidation.

Capital Stock Forms. *Capital stock* is issued to partners when the partnership is incorporated. These stocks are usually *common stock* and are assigned a *par value* (an arbitrary monetary value). Stock certificates are developed and issued to each partner based on his or her investment. These are then evidence of ownership. Capital stock accounts are controlling accounts, and each partner's stock balance is kept in a subsidiary account called a *stockholders ledger.*

COMMON STOCK ISSUE

EXAMPLE:

1. Articles of Partnership allow for issuing $50,000 of capital stock at $1 par value.
2. Partners' contributions:

	Contributing assets	Common stock issued (shares)
Partner A	$21,500	21,500
Partner B	11,800	11,800
Reserve (treasury)		16,700
Total		50,000

OPEN STOCK. Stocks may be sold to anyone outside the partnership, but this reduces the control of the partners since each share of stock represents a vote in the company's operation. If stocks are sold outside the partnership, subsidiary stockholders' ledgers are opened and often a single entry for *each* type of stock is made on the balance sheet and retained earnings statements.

PREFERRED STOCK. Another type of stock sometimes used is *preferred stock.* This type is almost like a bond in that a guaranteed stock dividend is stipulated. Each partner may have preferred stock. Dividends are paid out of retained earnings and these dividends are paid before common stock dividends. Preferred stock accounts are also maintained in separate subsidiary accounts for each partner.

Balance Sheet and Capital Statement Forms. For the partnership the balance sheet looks like that of the sole proprietorship except that the capital statement part shows each partner's capital, drawing account, and net income distribution separately if unequally shared.

For incorporated partnerships the individual entries are lost under retained earnings. However, there should be a note assigned to dividends paid to explain the distribution of these funds.

Corporation Forms

In this section, accounting for the corporation is illustrated. Recall that a corporation's capital is obtained through the selling of shares to stockholders. Therefore, a *retained earnings* statement is prepared in place of the capital statement used in an unincorporated enterprise. The balance sheet is more complex, as may be the income statement. In addition, some of the terms used on the forms change.

Income and Retained Earnings Statement Form. An example of such a statement is shown below.

ABC CONTRACTING CORPORATION
STATEMENT OF INCOME AND RETAINED EARNINGS
DECEMBER 31, 19X3

Contract operations	
Contract revenues earned	
Less contract costs	________
Gross profit	
Selling, general, and administrative	
Expenses	________
Income from operations	
Other income	
Gain on sale of equipment	
Interest expense (net of interest income)	
Income before taxes	
Provision for taxes (note 14)	
Net income (per share 10,000 shares at $2.753)	
Retained earnings beginning of year	________
Less dividend paid (per share $0.50)	________
Retained earnings, end of year	========

Balance Sheet Form. For corporations that have short-term contracts (those 1 year or less in duration) a standard or classified form of balance sheet is used. For corporations whose contracts are longer than 1 year, the classified form of balance sheet is not suitable; an unclassified one is better. The balance sheet below is an example of a classified or standard model.

CONTRACTING CORPORATION
BALANCE SHEET
DECEMBER 31, 19X3

Assets

Current assets

- Cash
- Certificates of deposits
- Money market certificates (note no.)
- Contract receivables
- Earnings on uncompleted contracts (note no.)
- Inventory
- Prepaid charges

Total current assets

Notes receivables (note no.)
Property and equipment net of
Accumulated depreciation (note no.)

Total assets

Liabilities and shareholders' equity

Current liabilities

- Current portion of long-term debts
- Accounts payable (note no.)
- Lease obligations payable
- Billings in excess of costs and estimated earnings on uncompleted contracts (note no.)
- Accrued income taxes
  - Current
  - Deferred
- Other current liabilities

Total current liabilities

Long-term debt (note no.)
Deferred income tax
Contingent liability (note no.)

Total liabilities

Shareholders' equity

Common stock, par value $1
Authorized 25,000 shares, issued and outstanding 10,000

Retained Earnings

Total shareholders' equity

Total liabilities and shareholders' equity

Notes

There are many notes cited on the balance sheet and income statements. These are an integral part of the statements. All types of contracting companies (unincorporated as well as incorporated) use them.

Consolidated Statements of Changes in Financial Position Form. As was explained in Chapter 2, managers need to know how company funds are being obtained (sources) and allocated (used). The *changes in financial position* statement shows this information. An example is shown below.

CONTRACTING COMPANY
CONSOLIDATED STATEMENTS OF CHANGES IN FINANCIAL POSITION
FOR YEARS ENDED DECEMBER 31, 19X3 and 19X2

	19X3	19X2
Sources of funds		
From operations		
Net income		
Charges to income (credit)		
Not involving cash		
Depreciation and amortization		
Deferred income tax		
Gain on sale of equipment		
Proceeds on equipment sold		
Net increase in billings on uncompleted contracts		
Decreases in inventory		
Decreases in prepaid charges and any other assets		
Increase in accounts payable		
Increase in other accrued liabilities		
Total		
Uses of funds		
Purchase of equipment		
Dividends paid		
Increase in contract receivables		
Increase in inventory		
Increase in notes receivable		
Increase in prepaid charges and and any other assets		
Decrease in accounts payable		
Decrease in notes payable		
Decrease in lease obligations		
Decrease in accrued income taxes payable		
Total		

(*continued*)

Increase (decrease) in cash and money market certificates (or CDs) for year		
Cash and money market certificates (or CDs)		
Beginning of year	________	________
End of year	$ ________	$ ________

Contract Receivables Form. This form may be single year or consolidated (2-year) type and explains the note on the balance sheet. An example of the consolidated type is shown below.

Contracts receivable	December 31, 19X3	December 31, 19X2
Contract receivables		
Billed		
Completed contracts		
Contracts in progress		
Retained		
Unbilled	________	________
Less allowance for doubtful accounts	________	________
Total contract receivables	$ ________	$ ________

Cost and Estimated Earnings on Uncompleted Contracts Form. This statement shows a summary of costs and estimated earnings on uncompleted contracts. Extensive background data collection is needed to present this form. Frequently, percentage-of-completion formulas are used to arrive at earnings, and sometimes estimated costs are used in lieu of actual costs. A single-year or a consolidated (comparative) statement may be used. An example of a consolidated statement is shown below.

	December 31, 19X3	December 31, 19X2
Costs incurred on uncompleted contracts		
Estimated earnings	________	________
Less: Billings to date	________	________
	$ ( )	$ ( )
Included in accompanying balance sheet:		
Costs and estimated earnings in excess of billings on uncompleted contracts		
Billings in excess of costs and estimated earnings on uncompleted contracts	$ ( )	$ ( )
	$ ( )	$ ( )

Property and Equipment Form. This form illustrates the note associated with *property* and *equipment* net of accumulated depreciation. An example is shown below.

	December 31, 19X3	December 31, 19X2
Assets		
Land		
Buildings		
Shop and construction equipment		
Trucks		
Leased equipment under capital leases		
Accumulated depreciation and amortization		
Buildings		
Shop and construction equipment		
Trucks		
Leased equipment		
Net property and equipment	$	$

Notes Payable Form. Sometimes it is desirable to explain the notes payable accounts. There probably are a variety of notes which were obtained under different loan conditions, and their payback periods are different. An example is shown below.

Notes payable	December 31, 19X3	December 31, 19X2
Unsecured notes (specified repayment schedule and interest payment)		
Secured notes (specified repayment schedule and interest payment		
	$	$

Income Taxes and Deferred Taxes Form. Visibility is enhanced relative to income taxes if this form is prepared. An example is shown below.

	December 31, 19X3	December 31, 19X2
Current payable, net of investment credits of $xxxx and $xxxx		
Deferred		
Contract related (current)		
Property and equipment related (current and noncurrent)		
Noncurrent		
	$	$

Earnings from Contracts Form. This is a multicolumn form largely because contracts usually will spread from one year to the next even if they are shorter than 1 year in duration. An example is shown below.

	19X3			19X2
	Revenues earned	Cost of revenues earned	Gross profit (loss)	Gross profit (loss)
Contracts completed during year				
No. Type				
Contracts in progress at year end				
No. Type				
Management contract fees earned				
Misc. contracts				
Unallocated costs chargeable to contracts		( )	( )	( )
	$	$	$	$

Other Forms. There may be a need for several other statement forms. Some are columnized; others are narrative in form.

1. Advances and equity in joint venture (columnized)
2. Accounts payable (narrative)
3. Leases (columnized)
4. Contingent liability (narrative)
5. Pension plans (narrative)
6. Contracts completed (columnized)
7. Contracts in progress (columnized)
8. Future work load (columnized)
9. Cash forecast (columnized)
10. Contract backlog and bidding information (columnized)

■ FORMS OF BUDGETS

Each contracting company's goal is to stay in business, earn a profit, and possibly grow to maturity. Proper budgeting is the managing of every part of the business, which may well spell the difference between success and failure. The accountant or accounting section should prepare budgets. These then act as (1) goals, (2) milestones, (3) controls over income and expense, and (4) allow for periodic assessments of how well the company is doing.

All budgets should be formally prepared. Budgets are sometimes called "forecasts." Finally, budgets are prepared for the short term (1 to 3) and the long term (5 to 10 years). We shall look at several short-term budgets first.

Contract Revenue Forecast Budget Form. This budget would list all possible sources of revenue projection for the next short term. Many of the data in this budget would come from the *schedule of contracts in progress* plus those projections that are reasonably secure. Some firms may set out specific dollar amounts for various types of building they wish to obtain. An example of a contract revenue forecast form is shown below.

CONTRACT REVENUE FORECAST ____ YEAR(s)

		Quarter				
Sources of revenue	Period:	1	2	3	4	Total
Contract no.						

Direct Materials Cost Budget Form. This budget is developed to show all *expected* direct materials costs. It may be summarized on a quarterly or other basis or may be made up according to sources of revenue (contracts). If inventories are to be maintained, this budget will show expected closing inventory as well as cost of inventory. An example is shown below.

DIRECT MATERIALS BUDGET ____ YEAR(s)

Contracts	
xxxx	
xxxx	
Desired inventory (year end)	________
Less estimated inventory January	________
Total direct materials purchased	________

Direct Labor Cost Budget Form. This budget provides projected direct labor cost figures. An example is shown below.

DIRECT LABOR COST BUDGET ____ YEAR (s)

Hours of direct labor	
Contract no. xxxx	
Contract no. xxxx	
Projected contracts	________
Total hours	
Hourly rate	$________
Total direct labor cost	$________

Contract Overhead Cost Budget Form. This budget itemizes the estimates that are charged against overhead. They include the following:

CONTRACT OVERHEAD COST BUDGET_____YEAR(s)

Indirect contract labor costs	
Indirect material costs	
Supervisory salaries	
Utilities	
Depreciation equipment	
Maintenance	
Security	
Insurance and property tax	__________
Other total contract overhead	__________

Cost of Contracts Sales Budget Form. This budget makes use of the direct materials purchase, direct labor costs, contract overhead cost budgets, and the estimated inventories. This budget will show the expected costs charged to contracts expected to be completed and those that will still be in process at year's end, taking into account material inventories. (For a residential contractor, this budget may be titled "Cost of houses sold.") An example is shown below.

COST OF CONTRACTS SALES BUDGET _____ YEAR(s)

Finished contracts			xxxx
Contracts in process		xxxx	
Less direct materials	xxxx		
Direct labor	xxxx		
Contract overhead	xxxx		
Total contract in progress cost		xxxx	
Less contracts in process inventory December		xxxx	xxxx
Cost of contracts finished			xxxx
Less contracts not sold December			xxxx
Cost of contracts sold			$xxxx

Operating Budget Form. The operating budget is another important one since it shows in summary form the expected balances from many general ledger accounts. Done properly, it also defines the percent represented by each part of the operation. Recall from Chapter 2 that when economic theory is applied, certain percentages are expected to remain constant year after year (or at least remain controlled). An example of an operating budget form is shown on the following page.

OPERATING BUDGET _____ YEAR(s)

Gross contract sales (revenue)	
Less: Cost of contract sales	______
Gross profit on contracts	
(percent of gross sales)	
Less: Selling expenses:	
(percent of gross sales)	
Less: General and administrative expenses	
(percent of gross sales)	______
Net operating profit	
(percent of gross sales)	
Other income	______
Less: Bad debts charged to completed jobs	______
Net profit before income tax	
(percent of gross sales)	
Provision for income tax	______
Net profit after income tax	
(percent of gross sales)	______

Note: Selling and general and administrative expense items would normally be listed.

Budget Income Form. With the budgets described above made as accurately as possible, the accountant would then make up the budget income statement. Its form would be similar to that of the company's annual income statement and should be prepared routinely.

Capital Expenditures Budget Form. The purpose of this budget is to show planned capital expenditures for the budget period. Usually, more than one year is included in this budget, an example of which is shown below.

CAPITAL EXPENDITURES BUDGET ___ YEAR(S)

Item	19x3	19x4	19x5	19x6
	$	$	$	$

Balance Sheet and Retained Earnings Budget Form. The entries on this form are exactly the same as those on the annual or periodic company balance sheet and retained earnings statements. However, the dollar amounts entered are predicted from all the other budgets and best estimates of managers and owners.

Cash Budget Form. This is a very important budget form because cash flow is frequently one of the most serious problems contractors must face—payrolls must be paid, for example. This budget form shows the short fall of cash and where loans will probably be required. An example is shown below.

CASH BUDGET YEAR

Estimated cash from:	Jan.	Feb.	Mar.	Apr.	etc.
Contract no.					
Contract no.					
Other	______				
Total cash receipts	______				
Estimated cash disbursements for:					
Contract costs					
Operating expenses					
Capital expenditures					
Other	______				
Total cash disbursements	______				
Cash balance beginning of month	______				
Cash balance end of month	______				
Minimum cash balance	______				
Excess or (deficiency)	$______				

Performance Budget Form. A series of performance budgets may be prepared by the accountant to illustrate the differences between the budgeted sums and actual monies spent and earned. These are valuable to managers. After studying them, managers can take action to prevent serious business failures, overspending, and so on.

REFERENCES AND SOURCES

1. Publications
   a. Collier, Keith, *Construction Estimating and Cost Accounting* (Englewood Cliffs, N.J.: Prentice-Hall, 1974).
   b. Wolkstein, Harry W., *Accounting Methods and Controls for the Construction Industry* (Englewood Cliffs, N.J.: Prentice-Hall, 1981).
   c. Adrian, James J., *Construction Accounting: Financial, Managerial, Auditing, and Tax* (Reston, Va.: Reston, 1979).
   d. "Building Contractors," in *Encyclopedia of Accounting Systems*, rev. ed., J. Pescow, Ed. (Englewood Cliffs, N.J.: Prentice-Hall, 1976).
   e. Lucas, Paul, D., *Accounting Guide for Construction Contractors* (Englewood Cliffs, N.J.: Prentice-Hall, 1973).

2. Associations and agencies
   a. American Institute of Certified Public Accountants, Inc. (AICPA), 1211 Avenue of the Americas, New York, NY. 10036
   b. Internal Revenue Service (various publications—see partial listing in Chapter 5).
   c. National Association of Home Builders, (NAHB), 15th and M Streets NW, Washington, DC 20005.
3. Educational sources
   a. College courses
      (1) Principles of accounting (four or five courses)
      (2) Accounting management
      (3) Financial management
      (4) Real estate principles
      (5) Computerized accounting
   b. Self-study—individual and group study materials from AICPA. (Subjects on accounting are liberally sampled. Each course includes one or more cassettes and a workbook. All courses are accredited.)

4

COMPUTER SERVICES

QUICK REFERENCES

More than ever, the cost of labor is going up; therefore, overhead costs are going up. To this add the various governments' demands that contractors collect more and more money for governmental purposes. This means one thing to all contractors: A management decision must be made to curb costs while continuing to operate and make a profit.

Most contractors can, if they will, analyze the variety of tasks done by employees that are routine and repetitive. These may include scheduling, mailing, accounting, inventory, and record keeping. A business computer can take over most of the tasks, thereby reducing their labor intensiveness. If personnel are retained, then efforts can be directed toward gaining additional business.

Each contractor should explore the value of owning a desktop business computer, or at the very least should consider leasing computer time.

DESKTOP MICROCOMPUTER

Each microcomputer is made up of a variety of components, each with a specific purpose. The various parts of the computer itself constitute the *hardware*. The *software* comprises the programs that direct the operation of the hardware.

Hardware

Microprocessor and CPU. The microprocessor (Figure 4-1) is the electronic heart of the computer. It makes possible the thousands of operations called for by the software programs. Microprocessors are usually included in the central processing unit (CPU). The CPU is the nerve center of the computer. It directs the actions and keeps track of everything.

Figure 4-1 Personal computer system. (Courtesy of Apple Computer, Inc.)

Memory. The computer's memory stores programs, information, and instructions temporarily until the data are ready for more permanent storage on disks or tapes.

Computer memory is usually measured in "K's." For example,

> 1K indicates a storage capacity of 1024 characters (each character = 8 bits or one byte).
>
> 64K indicates that approximately 64,000 characters or bytes can be stored at one time.

If a software program takes 20K of storage to operate a program and the user has a computer with a 64K memory, the operator can add only 44K of data at one time. Software designers provide needed storage information in their program descriptions. Memory sizes are cited as "RAM" (random access memory, e.g., 1K RAM, 64K RAM).

Keyboard. A keyboard needs to have all the characters of a typewriter, plus the number keys arranged as on an adding machine, plus control keys that allow the keyboard to interact with the computer. Many keyboards are connected by a cable. Data are entered into the computer through the keyboard (see Figure 4-1).

Monitor CRT. The monitor CRT resembles a television set, but usually is much clearer and easier on the eyes. Visual display is provided for all data being operated on by the operator.

Disks and Disk Drives. A disk is like a 45-rpm record or like a 78-rpm record minus the grooves. The smaller disks, called "floppy disks," are easy to handle and store very easily. Results of programs are stored on these magnetic disks. The larger disks, often called "hard disks," can hold much more information than the smaller "floppies." For example:

> A 5¼-inch floppy stores at 2K bytes (80 pages of type).
>
> A 5¼-inch double-density floppy can store to 640K bytes.
>
> A hard disk (about 8 inches) can store up to 20M bytes.

Disks need to rotate so that information can be written onto them and then read off. A disk-drive unit is needed for this job. Some programs require the use of two and even three disk drives.

Printers. Three types of printers provide output (hard copy) from the computer files.

1. *Thermal dot matrix printers* are fast, economical, and very reliable. They print by heating the paper when characters are to be printed. The quality is acceptable for drafts and interoffice copies.

2. *Impact dot matrix printers* form characters by striking the printer paper. They are very fast. Multiple copies can be made with these printers and the printing quality is some what better than that of the thermal printer. They cost more, too.
3. *Letter-quality impact printers* use a rotating plastic wheel called a "daisy wheel" to create characters on paper. These printers are much slower than either of the other two models, but their print quality is equal to that of the best typewriter. Type styles can be changed. The price is more than for the two other printers listed.

Ancillary Hardware and Equipment. Examples include:

1. Disk storage files and blank disks
2. Paper for printers, and paper catchers
3. Desk on which to set computer, and other furniture
4. Covers for protection
5. Software storage files
6. Interface cables, and cards
7. Interface with telephone
8. Language cards
9. Graphics table

Software

"Software" is a term that describes the programming that controls the internal functioning of the computer (system software) or the operating or (application) programming that the operator uses to perform the tasks needed.

Types of Programs

ROM PROGRAMS. Software programs built into the computer are called read-only-memory (ROM) programs. Floppy disks that contain packaged software programs such as those for accounting are ROM programs. These cannot be changed by the operator.

PACKAGED PROGRAMS. There are well over 2000 different packaged programs. These are sometimes listed as "off-the-shelf" programs. Prices range from $50 to $1500.

Some packaged programs may have the capacity to be modified. Usually, minor modifications are permitted.

CUSTOM PROGRAMS. If no program satisfies the need of a contractor, software designers will develop a program to meet the need. These are very expensive.

Software Categories

FINANCIAL MODELING/PLANNING. *Financial modeling/planning* programs aid contractors to improve their operations and profit outlook in a variety of ways. Examples include:

"Business Planner"
"Loan Analysis"
"Construction Cost/Profit Analysis"
"Financial Management System"
"Depreciation"
"Project Control"
"Statement of Net Worth"

GENERAL ACCOUNTING. *General accounting* programs aid contractors by taking over the routine tasks associated with accounting. Examples include:

"Accounts Receivable/Payables"
"Business Accounting Package"
"Construction Job"
"Construction Bookkeeper"
"General Accounting System"
"Performance Manager"
"Ledgers—Various Kinds"

INVENTORY CONTROL. *Inventory control* programs aid contractors who warehouse supplies and those who purchase only enough for each job. Examples include:

"Inventory Control"
"Inventory Manager"
"Inventory Program"

JOB CONTROL. *Job control* programs aid the contractor by maintaining up-to-the-minute status on each job in progress. Examples include:

"Job Control Program"
"Job Cost System"
"Job Scheduling"
"Job Costing Model"

PAYROLL. *Payroll* programs not only keep statistics on each employee but provide data for automatic paycheck writing and compile tax data. Examples include:

"Paymaster"
"Payroll"
"Pay System Accountant"
"Payroll—Personnel, Cost Distribution, etc."

REAL ESTATE. *Real estate* programs may aid contractors if these contractors deal frequently in real estate. Examples include:

"Property Management"
"Real Estate Analysis"
"Rent versus Buy"
"Tax-Deferred Module"

TAXES. *Tax* programs reduce the drudgery involved in preparing monthly, quarterly, and annual taxes, and aid in planning and preparation. Examples include:

"Individual Tax Plan"
"Personal Tax Plan"
"Tax Planner"
"Tax Preparer"
"Tax Manager"

ENGINEERING. Almost any engineering problem—architectural, civil, electrical, and so on—is available as software. Examples include:

"Beams"
"Drawing List"
"Engineering Math"
"Liquid and Gas Flow Calculating"
"Production Control"
"Quality Control Group"
"Solar Energy"
"Surveyor's Guide"
"T-Square"
"Wind-Energy Calculating"

MISCELLANEOUS. Many programs very useful to contractors fall in the miscellaneous category. Examples include:

"Cash Master"
"Checkwriter"
"Construction Materials List Generator"
"Depreciation"
"Critical Path/Profit Analysis"
"Office Manager"
"Production Manager"
"Project Planning"
"Project Scheduling"
"The Estimator"
"The Price Quoter"
"Time and Money Meter"

Sources of Software. Almost every microcomputer retailer has a large variety of software listings.

1. Apple dealers (see the phone book)
2. Radio Shack outlets (see the phone book)
3. IBM retailers (see the phone book)
4. Commodore business machines dealers (see the phone book)
5. Burroughs (see the phone book)
6. Tiny Systems, Inc., Richardson, TX 75081
7. Advanced Software Technology, Inc., Overland Park, KS 66204

ADVANTAGES OF OWNING A COMPUTER

"User Friendly." Probably the most important factor to a contractor who owns a computer is that the computer be designed to be *"user friendly."* This means that the machine is easy to use and that the software is designed so that all the assistance an operator needs is provided in the program. As decisions or data are needed by the program, the monitor prints a series of questions. Responses to the question are operator inputs.

Monitoring. The computer can be used to *monitor all phases* of each profit center (job or project). It can *consolidate like elements* across profit centers to give year-to-day totals for *incomes* and *expenses.* It can monitor year-to-day *overhead* and any other information that aids in financial or accounting management.

Modeling. Many, many models of software programs are available that aid the contractor in *decision making.* Most of the management and economic concerns discussed in Chapters 2 and 3 can be modeled. *Estimtating, job* *costing*, *production control*, and *inventory control* are just a few of the available models.

Accounting. All aspects of accounting are handled easily by computers. Especially important would be *output reports* of *comparative analysis*, *sources and uses of working capital*, *capital planning*, and *cash flow analysis*.

Office Efficiency. Office personnel at the *base* or in the *field* can make extensive use of the computer to *file data*, *plans*, *specifications*, *estimates*, *mailing lists*, and for a host of other uses. With a phone patch and a modem (modulator-demodulator), field office personnal have ready access to any and all data related to their profit center. This can save many hours and dollars in travel and other expenses.

Graphics. With a graphic generator and associated software, all manner of *construction problems can be analyzed* quickly and accurately. The possible results could include considerable savings in materials and labor.

TECHNIQUES TO USE WHEN BUYING A MICROCOMPUTER

Definition

All contractors should work with a reputable microcomputer dealer. But before he or she contacts a dealer, a plan should be drawn up to aid the dealer in meeting the contractor's needs. This is where the definition becomes important. The definition needs to include a narrative of the following items.

1. Pinpoint all the applications in the office and at the site that are expected to be handled by the computer. Examples include:
   a. Accounting
   b. Job planning, estimating
   c. Data base management
   d. Scheduling
   e. Job cost accounting
   f. Engineering
2. Define those applications that will be cross-linked. Examples include:
   a. Accounting and job cost accounting
   b. Job planning and bidding
   c. Data base management and job cost centers
   d. Scheduling and job cost accounting
   e. Field office-to-base office communication

3. Define the extent (volume) of entries per application. Here the software plays an important roll since each software program identifies how much RAM is needed to run or operate the program and how much data it is expected to handle. For example:
   a. For each company or person on a mailing list (data usually included on a 3 by 5 card) 180 to 200 typewriter keystrokes are needed. Therefore, if a file contained 50 such cards, the computer disk storage would be 50 × 200 = 10,000 characters.
   b. An annual budget could easily use up 20K memory, plus 30K for the program.
   c. Word processing uses a lot of memory. Ten pages of a single-spaced report would require about 25K plus the 30K for the operating program.
4. Define the order of importance for each family of applications if ordering is important. For example:
   a. Accounting is always ordered.
   b. Data base management is ordered.
   c. Job production and costs are ordered.
   d. Profit center records are ordered.
   e. Output reports are ordered.

Job Description

Having defined all applications, prepare a job description for each application. Make each description clear and definitive, but complete. The dealer will attempt to define your needs based on the various descriptions. Should they be vague or sketchy, he or she may oversell to preclude making errors in interpretation.

The descriptions have a twofold purpose: (1) they aid in defining hardware needs, and (2) they aid in the selection of software programs.

Assembling the Cost of a Computer System

Use the following list or make a similar list. Make sure that it includes the following main categories:

1. The computer, keyboard, monitor, and disk drive(s)
2. Peripheral equipment and accessories, such as:
   a. Printer, modem, furniture, floppy disks
   b. Graphic table, paper catcher, interface cards
3. Software programs
4. Supplies, paper, disks, disk storage files
5. Training costs
6. Service contract costs
7. Insurance costs

Total estimated cost $________

TERMS AND DEFINITIONS

ASCII: American Standard Code for Information Interchange; computers use binary numbers (combinations of ones and zeros) to represent letters, numbers, and special characters; the ASCII code specifies which binary number will stand for each character and provides a standard that allows computers from different manufacturers to "talk" to each other.

BASIC: Beginners All-purpose Symbolic Instruction Code; the most popular language for personal computers.

Baud: A measure of the speed at which computer information travels (normally, between a computer and a peripheral or between two computers). A baud is equal to the number of bits per second.

Byte: The basic unit of measure of a computer's memory, representing a single alphanumeric character (A, B, C, 1, 2, ?, &, etc.); the equivalent of 8 bits.

COBOL: COmmon Business-Oriented Language; a computer language developed specifically for business use.

Computer Program: A series of commands, instructions, or statements put together in a way that tells a computer to perform a specific task or series of tasks.

CPU: Central processing unit; the part of the computer that collects, decodes, and executes instructions; often made up of a microprocessor and associated circuitry.

Data Base: A collection of related information, such as in a mailing list, that can be sorted in the computer and retrieved in several ways.

Disk: A flat, circular device that resembles a phonograph record and is used to store information and programs for the computer. Two types of disks are used with personal computers: floppy disks (or diskettes) and hard disks.

Disk Drive: The machinery that operates either a floppy or a hard disk, rotating it at high speeds to read information stored on the disk or to write new information on it.

Floppy Disk: A flexible plastic disk, enclosed in a cardboard jacket, that stores information generated by the computer; in small computers, usually either 5¼ or 8 inches in diameter.

FORTRAN: FORmula TRANslation, a high-level computer language used primarily for mathematical computations; although FORTRAN is available for some small computers, it is used primarily with large computer systems.

Hard Copy: Printed information generated by a computer.

Hard Disk: A magnetically coated metal disk, usually permanently mounted within a disk drive; capable of storing 30 to 150 times more information than can a floppy disk; also called a Winchester disk.

High-Level Language: A programming language that allows a person to give instructions to a computer in English rather than in the numerical (binary) code of ones and zeros; BASIC, COBOL, and Pascal are examples of high-level languages.

Input/Output (I/O): Software or hardware that exchanges data with the outside world.

Language: A code that the computer understands; low-level languages resemble the fundamental codes of the computer; high-level languages (such as BASIC and COBOL) resemble English.

Memory: The part of the computer that stores program instructions and data.

Microprocessor: Core of the central processing unit of a computer; it holds all the essential elements for manipulating data and performing arithmetic calculations.

Modem: MOdulator-DEModulator; a device that converts a computer's electrical signals into audible sounds (modulation) for transmission over the telephone, and back again (demodulation) for reception via telephone, used to link one computer to another.

Operating System: A group of programs that act as intermediaries between the computer and the application software; the operating system takes a program's commands and passes them down to the central processing unit in a language that the CPU understands; application programs must be written for a specific operating system, such as DOA, SOS, CP/M, and others.

Pascal: A high-level programming language with a larger, more sophisticated vocabulary than BASIC, used for complex applications in business, science, and education; named after the seventeenth-century French mathematician.

Program: A sequence of instructions that the computer can understand and execute.

RAM: Random access memory; the part of the computer's memory that allows both reading and writing, as opposed to ROM.

ROM: Read-only memory; the part of the computer's memory that allows just reading and is used to hold information that never changes, such as a computer language.

Save: To store a program on a disk or somewhere other than in the computer's memory.

Software: Instructions that operate the computer hardware; system software is used for the computer's general tasks or functions (such as operating a printer or understanding BASIC); application software is used to accomplish a specific task, such as financial modeling or word processing.

Terminal: A piece of equipment used to communicate with a computer, such as a keyboard for input, or a video monitor or printer for output.

Video Monitor (CRT): The computer picture screen; also called simply the monitor.

Word Processing: The entry, manipulation, editing, and storage of text using a computer.

REFERENCES AND SOURCES

1. Educational sources
   a. Junior, senior, and business colleges offer weekend courses in microcomputer. Usually, more than one computer dealer participates.
   b. Junior-college-level courses:
      (1) Data processing
      (2) BASIC
      (3) COBOL
      (4) FORTRAN
      (5) Keypunching
   c. Business schools:
      (1) Computer fundamentals
      (2) Computer operations
      (3) Computer systems
      (4) Microcomputers
      (5) Software development
   d. Correspondence schools offer courses that usually include a computer together with literature and lessons.
   e. Retail dealers conduct education/training programs. These usually are made a part of the original sales contract. Some dealers offer as little as 1 hour of instruction, others as much as a week.

2. Literature
   a. *Personal Computer in Business*, Apple Computer
   b. *Applications Source Book*, Radio Shack

3. Periodicals
   a. *Mini-Micro Systems*
   b. *Journal of System Management*
   c. *Datamation*
   d. *Illustrated*
   e. *Info Systems*
   f. *Journal of Operations Research*
   g. *Editor and Publisher*
   h. *Electronics N*
   i. *Electronics World*

5

LEGAL REQUIREMENTS AND CONSIDERATIONS

QUICK REFERENCES

All contractors conducting business in the United States are legally required to meet a variety of rules. They must protect themselves. The subjects in this chapter identify many of the requirements to be met and also include some considerations that protect the contractor. When referencing any of the subjects listed above, remember that compliance in almost every circumstance makes it possible for a contractor to seek relief through the courts.

SOLE PROPRIETORSHIP

A *sole proprietorship* is the simplest form of a contracting organization. The contractor is the sole owner. The business liabilities are the contractor's personal liabilities. The proprietary interest dies when the contractor dies. Because the contractor takes all the risks, he or she is entitled to all the profits (less tax obligations) but is simultaneously obligated to risking the business's assets and *all personally owned assets* (except those exempted by legal state statutes).

Parts of the Company

1. Identifying the company by name, advertising, etc.
2. Registering the company through licensing
3. Adopting an accepted form of record keeping
4. Assets
5. Real property and depreciable property
6. Accounts and notes receivable
7. Merchandise inventories
8. Land and leaseholds
9. Buildings, machinery, furniture and fixtures
10. Patents, copyrights
11. Goodwill
12. Agreement not to compete for a fixed period of time
13. Professional skill of the contractor or his or her company
14. Opportunity to dispose of the company

PARTNERSHIP

A *partnership* is the formal relationship between two or more persons who join together to carry on a construction business, with each person contributing money, property, labor, or skill and each expecting to share the profits and losses of the contracts into which they enter.

Joint Venture

Contractors who participate in joint ventures to share the risks, and to combine their financial and other resources as well as their talents, are considered legally to be partners to the extent of the venture. Profits and losses are shared as partners share. Liabilities are equally chargeable to each contractor in the joint venture. If any contractor is unincorporated, his or her liability is that of a sole proprietorship. If any are incorporated, their liability extends to the assets of the company (unless otherwise committed).

Uniform Partnership Act

This is an act that has been adopted by 47 states. Those not using the act are: Georgia, Louisiana, Mississippi, the District of Columbia, the Virgin Islands, and Guam. There are seven parts to the act:

I. Preliminary provisions
II. Nature of partnership
III. Relations of partners to persons dealing with the partnership
IV. Relations of partners to one another
V. Property rights of a partner
VI. Dissolution and winding up
VII. Miscellaneous provisions

The purpose of the act is to make uniform the law of partnership. This act contains rules, terms, and descriptions regarding every facet of establishing and operating a partnership. All persons considering forming a partnership should study this act thoroughly.

Formation of a Partnership

The following outline of the *formation of a partnership* provides only the basic or bare-bones structure with some explanation. A basic partnership agreement should cover the following points:

1. Name and business
2. Term
3. Capital
4. Profit and loss
5. Salaries and drawings
6. Interest on capital
7. Management, duties, and restrictions
8. Banking
9. Books
10. Voluntary liquidation
11. Retirement
12. Death
13. Arbitration
14. Signature blocks

Of particular importance, legally, are the clear definitions of a fiscal year, retirement considerations, and rights to property upon the death of one partner.

Partnership Form with Outlines of Parts

Opening Paragraph. Agreement statements:

1. Date
2. Names of partners
3. Each partner's address

Name of Business. Basic data:

1. Name of company
2. Purpose of company
3. Principal place of business

Term of Business. Statement of term or duration of partnership:

1. Beginning date of partnership
2. Termination date or continuing statement

Capital. Capital investments and accounting statement description:

1. Partner A's capital investment
2. Partner B's capital investment
3. Define capital accounts
4. Restrictions on use of capital

Profit and Loss. Profit-and-loss distribution:

1. Stated division of profits or losses
2. Drawing accounts are defined (see Chapter 3 for definition)
3. Tax liability defined

Salaries and Drawings. Statement of salaries and drawings:

1. Restrictions and limitations are specified
2. Drawing account limitations are cited

Interest. Statement specifying the interest on capital investments:

1. Limitations to application of interest on capital investments
2. Detailed description on interest application

Management, Duties, and Responsibilities. Narrative of management, duties, and responsibilities of each partner:

1. Rights of each partner
2. Percent of each partner's time allocated to business
3. Restrictions on involving company in business or outside interests
4. Restrictions on borrowing, lending, and leasing
5. Restrictions on assigning, mortgaging, and selling assets
6. Restrictions of illegal acts

Banking. Statement specifying the types of banking and uses of banking:

1. Use of checking accounts mandatory
2. Signature controls set in writing
3. Withdrawal procedures established

Books. Accounting books and related accounts of the company:

1. Maintained in accordance with sound business practices
2. Audit cycle stated
3. Maintained at a specific location
4. Kept on a fiscal basis

Voluntary Termination. A narrative of the conditions and process of termination:

1. An agreement statement to dissolve the partnership
2. Times specified
3. Selling of partnership's name and assets conditions
4. Use of assets to liquidate liabilities and obligations of the company
   a. Pay expenses, accounts payable, notes payable
   b. Equalize income accounts of partners
   c. Discharge balances in income accounts
   d. Equalize and discharge capital accounts

Retirement. A narrative setting policy and conditions for a partner to retire:

1. Time of retirement is specified
2. Written notice is given
3. Rights of remaining partner to purchase interest of retiring partner, or
4. Right to terminate the partnership
5. Notice to do point 3 or 4 above is made in writing
6. Conditions of establishing purchase price and payments

Death. A statement to describe activities when one of the partners dies:

1. States the rights of surviving partner to liquidate or purchase interest of decedent
2. All decisions must be in writing in a specified time period
3. Valuation of decedent's accounts establishes each partner's share and purchase price
4. Restriction on allowances may include:
   a. Goodwill
   b. Trade names
   c. Patents
   d. Intangible assets
   e. Interest rates where purchase price is made in installments

Arbitration. A statement of policy on arbitration:

1. Sets forth rules of arbitration
2. May use American Arbitration Association
3. Subjects suitable for arbitration
4. Subjects not allowed in arbitration

Signature Blocks. A swearing statement is written followed by signature blocks:

"In witness whereof the parties have signed this agreement"

____
Partner A's name

____
Partner B's name

Sources of Acts

1. Secretary of state of the state in which you are planning to do business
2. Federal laws and regulations (FCC) Washington, DC
3. Uniform laws or acts
4. Code of federal regulations
5. Model Act (model business corporation act)

6. *West's Business Law* (St. Paul, Minn.: West Publishing)
7. Bender's Uniform Commercial Code Service
8. *Martindale Hubbell Law Dictionary*, Law Digest of Uniform Acts (Summit, NJ: Martindale Hubbell, Inc.).

Places to Obtain Acts

1. Public library
2. Law library
3. Lawyer's office library
4. Federal or state government offices

ARTICLES OF INCORPORATION

This section provides data on the articles of incorporating a company with specifications, filing the articles with the secretary of state, and the process to be followed for developing stocks and bonds.

Minimum Articles for Incorporation

The basic and minimum articles needed to incorporate are:

1. Name
2. Principal office and registered agent
3. Purpose or purposes
4. Capital stock
5. Incorporators
6. Number of classes of directors
7. Management
8. Compromise or arrangements with creditors
9. Certification and signature blocks

Name. A simple statement of the name of the corporation is listed under this heading.

Principal Office and Registered Agent. A descriptive statement that includes the following:

1. The location of the principal office of the corporation
   a. Street, city
   b. County and state
2. The registered agent's name at the address above

Purpose(s). A detailed statement that includes *all* activities in which the business will be legally licensed to engage:

1. To conduct or promote various activities
   a. Construct, buy, sell, design, own, use, lease, render service, etc.
   b. To engage in any lawful act or activity
2. To act in accordance with the general corporation law of the state to which the company is licensed

Capital Stock. This is a statement that identifies:

1. The total number of shares of stock the corporation will be allowed to issue
2. The type of stock(s) (common or preferred) that will be issued
3. The par value of the stock

Incorporators. Under this heading all the incorporators and their current mailing addresses are listed.

Number and Classes of Directors. This section will either specify the exact number and class of director or may make reference to the corporation bylaws. The bylaws usually detail the makeup, duties, and service of the directors. Special initial appointments and other such instructions may be a part of this heading. Some may include:

1. Increasing and decreasing their number
2. How elected
3. Classification of additional directors
4. Election at annual stockholders meeting

Management. This is the longest article of the incorporation form. It usually specifies in exact detail the things that the board of directors can and cannot do.

1. Hold meetings
2. Make, alter, and repeal bylaws
3. Set policy for open review of the corporation books by stockholders
4. Declare and pay dividends
5. Establish reserves for working capital or to reduce reserves
6. Make lawful changes to the capital accounts
7. Use or apply funds to treasury stock
8. Set policy to limit stock ownership of directors, officers and employees of the corporation

9. Establish health and accident benefit programs
10. Establish a retirement plan and allow for its funding
11. Issue obligations (bonds), both secured and unsecured, and set forth their conditions
    a. Types of bonds are usually listed
    b. Types of mortgages are also listed
    c. Pledging security by type may be included

Compromise or Arrangements with Creditors. This section usually details the conditions that state how compromises or arrangements will be handled. It also details which title and section of the state's code to follow.

This section may also include how reorganization of the corporation is to be carried out, if needed.

Certification and Signature Blocks

"In witness whereof the undersigned, being all of the incorporators herein before named, do hereby make this certification for the purpose of forming a corporation pursuant to the general law of the state of ( ), and do hereby certify that the facts herein before set forth are true and correct and have accordingly here unto set our hand and seals this (month, day, year).

____
Incorporator #1

____________________ Signature blocks
Incorporator #2

____
Incorporator #3

(Acknowledgments)

Sample Set of Forms of Incorporation

A sample set of incorporation forms is shown in Figure 5-1.

STATE OF MISSISSIPPI

SECRETARY OF STATE

EDWIN LLOYD PITTMAN
POST OFFICE BOX 136
JACKSON, MISSISSIPPI 39205
TELEPHONE (601) 354-6541

ASSISTANT SECRETARIES:
R.M. ARENTSON, JR.
SECURITIES AND ADMINISTRATION
RAY BAILEY
CORPORATIONS AND UCC
A. MICHAEL ESPY
PUBLIC LANDS

DIVISION DIRECTORS:
ALBERT D. EAST
ACCOUNTING
THOMAS H. ELLIS
PERSONNEL

We are enclosing two copies of the articles of incorporation form to be executed by a business desiring to incorporate under the Mississippi Business Corporation Act. The documents must be completed in all details including street or road names in Articles Seventh, Eighth and Ninth and forwarded to this office in duplicate with the proper fee. We must be in receipt of two copies of the articles with original signatures properly notarized in order to process the document.

The fee for filing the articles of incorporation in this state and issuing a certificate of incorporation is on a basis of $25.00 for the first five thousand dollars of authorized capital and $2.00 for each additional thousand dollars of authorized capital and $2.00 for each additional thousand dollars or part thereof of authorized capital stock. Authorized capital is figured to be the number of authorized shares of stock multiplied by the par value per share or sales price per share. Authorized capital exceeding $250,000.00, no par value stock or stock with par and no par value require the maximum filing fee of $500.00.

If you have any questions, please feel free to contact the Corporate Division of this office.

Very truly yours,

EDWIN LLOYD PITTMAN
SECRETARY OF STATE

By:

Ray Bailey
Assistant Secretary of State

RB:lsl

Figure 5-1 Sample articles of incorporation. *Note*: Libraries in each town would usually have these types of applications/forms for their respective states. (Courtesy of Gulfport, MS, Law Library) (*continued*)

(TO BE EXECUTED IN DUPLICATE)
ARTICLES OF INCORPORATION
OF

____

We, the undersigned natural persons of the age of twenty-one years or more, acting as incorporators of a corporation under the Mississippi Business Corporation Law, adopted the following Articles of Incorporation for such corporation:

FIRST: The name of the corporation is ____

____

SECOND: The period of its duration is ____
(May not exceed 99 years)

THIRD: The specific purpose or purposes for which the corporation is organized stated in general terms are:

(It is not necessary to set forth in the Articles of Incorporation any of the powers set fourth in section 79-3-7 of the Mississippi Code of 1972.)

(Use the following if the shares are to consist of one class only.)

FOURTH: The aggregate number of shares which the corporation shall have authority to issue is ________ ______________of the par value of____________Dollars ($______) each (or without par value)
(par value or sales price shall not be less than $1.00 per share) (If no par shares are set out, then the sales price per share, if desired.)

(Use the following if the shares are divided into classes.)

FOURTH: The aggregate number of shares which the corporation is authorized to issue is____________ divided into____________ classes. The designation of each class, the number of shares of each class and the par value, if any, of the shares of each class, or a statement that the shares of any class are without par value, are as follows:

Number of Shares	Class	Series (if any)	Par Value per Share or Statement That Shares are Without Par Value

Figure 5-1 *(continued)*

The preferences, limitations and relative rights in respect of the shares of each class and the variations in the relative rights and preferences as between series of any preferred or special class in series are as follows: (Insert a statement of any authority to be vested in the board of directors to establish series and fix and determine the variations in the relative rights and preferences as between series.)

FIFTH: The corporation will not commence business until consideration of the value of at least $1,000 has been received for the issuance of shares.

SIXTH: Provisions granting to shareholders the preemptive right to acquire additional or treasury shares of the corporation are:

SEVENTH: The street and post office address of its initial registered office is ______________________

____

(Street and Number) (City) (State)

and the name of its initial registered agent at such address is ______________________________

EIGHTH: The number of directors constituting the initial board of directors of the corporation, which must be not less than three (3), is ______________ and the names and addresses of the persons who are to serve as directors until the first annual meeting of shareholders or until their successors are elected and shall qualify are:

NAME	STREET AND POST OFFICE ADDRESS
______________________	______________________
______________________	______________________
______________________	______________________

Figure 5-1 *(continued)*

NINTH: The name and post office address of each incorporator is:

NAME	STREET AND POST OFFICE ADDRESS
______________________	______________________
______________________	______________________
______________________	______________________

TENTH: (Here set forth any provision, not inconsistent with law, which is desired to be set forth in the Articles: Including, any provision restricting the transfers of shares or any provision required or permitted to be set forth in the by-laws)

Dated ______________ 19 ____ ______________________

____

____

Incorporators

ACKNOWLEDGMENT

STATE OF MISSISSIPPI

County of ______________________

This day personally appeared before me, the undersigned authority ______________________

______________, ______________, ______________.

incorporators of the corporation known as the ______________________
who acknowledged that they signed and executed the above and foregoing articles of incorporation as their act and deed on this the ________ day of ______________________ 19____

____

Notary Public

My Commission expires ______________________
(NOTARIAL SEAL)

Note: On all addresses the street and number must be shown if there is a street or number.

Figure 5-1 *(continued)*

SUBCHAPTER S CORPORATION

A *Subchapter S corporation* is a type of corporation in which the stockholders elect to be taxed in a manner similar to partners rather than allowing the corporation to pay income taxes.

Eligibility Requirements for Election to Subchapter S

1. The corporation must be domestic.
2. It must *not* be a member of an affiliated group.
3. It must have only one class of stock (usually common).
4. It must *not* have more than 25 shareholders prior to 1982, 35 members after 1982.
5. It must include only individuals, estates, or certain trusts.
6. It must *not* have a nonresident alien as a shareholder.

Advantages

1. All corporate benefits are available to Subchapter S corporations.
2. Shareholders may work for the corporation and the corporation withholds federal and state taxes due.
3. All stockholders share in all the dividends (net earnings) whether distributed or not.
4. Retirement plans may be established.
5. Stock bonus or profit-sharing plans are possible.
6. Shareholders share in the corporations excess net long-term capital gain.
7. Excess of capital loss is not passed onto the shareholders.
8. A corporate net operating loss is treated by the shareholder as a business loss (Schedule C of 1040).

CONTRACTS AND CONTRACT FORMS

Several important subjects are covered in this section. All contractors must fully understand what a contract is and what it means to the customer. The subjects include a legal definition of a contract, purposes for contracts, elements of legal contracts, types of contracts, and the contracting process.

Legal Definition of a Contract

An agreement between two or more persons which creates an obligation to do or not do a thing.

Construction Contract. A type of contract in which a set of plans and specifications are made a part of the contract; one that customarily is secured by performance and payment bonds. This bonding protects both subcontractors and the party who has contracted to have the work done.

Subcontract. A contract that is subordinate to another contract. For example, a general contractor contracts to build a house for Mr. X; the contractor then issues a subcontract to the electrician for installing the electrical service in the house.

Purposes of Contracts

1. Engineering services
2. Design services
3. Sales
4. Real estate
5. Equipment lease-purchase
6. Architectural services
7. Preconstruction
8. Construction
9. Materials
10. Labor
11. Trade
12. Construction loan
13. Appraisal
14. Building permits
15. Insurance

Elements of a Contract

For a contract to withstand a test by law, it must have several parts clearly specified. Other parts may be added to improve the binding of the parties as well as to specify the intent more clearly.

Competent Parties. Both parties entering into a contract must have legal capacity. This means that both must be of legal age and must be legally free to enter into the contract. Being legally free means that one is not a sentenced criminal, lunatic, infant, or minor. Minors may enter into contracts legally for certain necessities (food, shelter, and clothing).

Lawful Subject or Object. A second element needed in a contract is a lawful purpose. Contracting for a house would be lawful subject or object.

This element is usually satisfied by describing it and citing its geographical location. Contracting to build some project intended for illegal purposes is an invalid contract and has already been tested in court.

Legal Consideration. *Both parties must receive legal consideration*, and these considerations must be specified in the contract. In a construction contract the builder usually expects payment in money for work performed; that is his or her consideration. The owner, on the other hand, obtains, for example, a completed house or building as his or her consideration. A contract is invalid if only one party receives consideration.

Mutuality of Agreement. A mutual agreement element of a contract specifies the conditions under which the contractor submits his or her bid and the conditions under which the owner accepts the bid.

1. The offer and its acceptance must be unqualified—there must be no terms or conditions.
2. With construction contracts the bidding documents usually constitute agreement and form mutual agreement when accepted.
3. Times, turn-key dates, payment schedules, and other schedules are frequently elements of mutual agreements.
4. Quality standards may also become elements of mutual agreements.

Mutuality of Obligation or Genuine Intent. This element of the contract is extremely serious and implies that a genuine (not fraudulent) offer is made by a contractor to engage in the project *and* that acceptance has equally genuinely been made by the owner. In other words, both parties must truly and willingly intend to enter into the contract. Either party must always be cautious and be on the lookout for invalidation, which may come from:

1. Duress by one party by threat of violence
2. A mistake in regard to subject
3. Undue influence by outside parties
4. Misrepresentation of the facts (but without intending to do so)
5. Fraud by one of the parties

Types of Construction Contracts

Types of contracts fall into four broad areas: fixed price or lump sum, time and material, cost types, and unit price. Each of these general broad categories is defined further below.

Fixed-Price Contracts. A fixed-price or lump-sum contract is a contract in which the price is not usually subject to adjustment because of costs incurred by the contractor. Common variations of fixed-price contracts are:

1. *Firm fixed-price contract:* A contract in which the price is not subject to any adjustment by reason of the cost experience of the contractor or his performance under the contract.

2. *Fixed-price contract with economic price adjustment:* A contract that provides for upward or downward revision of contract price upon the occurrence of specifically defined contingencies, such as increases or decreases in material prices or labor wage rates.

3. *Fixed-price contract providing for prospective periodic redetermination of price:* A contract that provides a firm fixed price for an initial number of unit deliveries or for an initial period of performance and for prospective price redeterminations either upward or downward at stated intervals during the remaining period of performance under the contract.

4. *Fixed-price contract providing for retroactive redetermination of price:* A contract that provides for a ceiling price and retroactive price redetermination (within the ceiling price) after the completion of the contract, based on costs incurred, with consideration being given to management ingenuity and effectiveness during performance.

5. *Fixed-price contract providing for firm target cost incentives:* A contract that provides at the outset for a firm target cost, a firm target profit, a price ceiling (but not a profit ceiling or floor), and a formula (based on the relationship that final negotiated total cost bears to total target cost) for establishing final profit and price.

6. *Fixed-price contract providing for successive target cost incentives:* A contract that provides at the outset for an initial target cost, an initial target profit, a price ceiling, a formula for subsequently fixing the firm target profit (within a ceiling and a floor established along with the formula, at the outset), and a production point at which the formula will be applied.

7. *Fixed-price contract providing for performance incentives:* A contract that incorporates an incentive to the contractor to surpass stated performance targets by providing for increases in the profit to the extent that such targets are surpassed and for decreases to the extent that such targets are not met.

8. *Fixed-price level-of-effort term contract:* A contract that usually calls for investigation or study in a specific research and development area. It obligates the contractor to devote a specified level of effort over a stated period of time for a fixed dollar amount.[1]

Each of these types of contracts has advantages to both the contractor and the owner, and both have unique disadvantages as well. Both parties would do well to study the various types of contracts before selecting one. It is almost universally true that when one party gains, the other party gives something up.

[1] *AICPA Industry Audit Guide, Audits of Government Contractors* (New York: American Institute of Certified Public Accountants, 1975), pp. 3-4.

Time-and-Material Contracts. Two basic elements of the contract are used to determine the contract costs. The contract may be paid based on direct labor hours alone, or on direct labor hours and material cost (direct and indirect costs and profit are included in the pricing). Several variations are:

1. *Time only at a marked-up rate:* The rate is established to cover variable and fixed costs attributable to the contract, and profit, but not usually material costs. Materials may be provided by the owner.
2. *Time at a marked-up rate; materials at cost:* When using this contract the contractor usually sets the rates so that each DLH contracted for includes direct labor, variable overhead, fixed costs, and profit.
3. *Time and material at marked-up rates:* The markup rates may be different for time and materials.
4. *Guaranteed maximum cost:* Type 1 or 3 is used.

Cost-Type Contracts. Cost-type contracts provide for reimbursement of allowable or otherwise defined costs incurred plus a fee that represents profit. Cost-type contracts usually require only that the contractor use his best efforts to accomplish the scope of the work within a specified time and stated dollar limitation. Common variations of cost-plus contracts are:

1. *Cost-sharing contract:* A contract under which the contractor is reimbursed only for an agreed portion of costs and under which no provision is made for a fee.

2. *Cost-without-fee contract:* A contract under which the contractor is reimbursed for costs with no provision for a fee.

3. *Cost-plus-fixed-fee contract:* A contract under which the contractor is reimbursed for costs plus the provision for a fixed fee.

4. *Cost-plus-award-fee contract:* A contract under which the contractor is reimbursed for costs plus a fee consisting of two parts: (a) a fixed amount that does not vary with performance, and (b) an award amount based on performance in areas such as quality, timeliness, ingenuity, and cost-effectiveness. The amount of award fee is based on a subjective evaluation by the government of the contractor's performance judged in light of criteria set forth in the contract.

5. *Cost-plus-incentive-fee contract (incentive based on cost):* A contract under which the contractor is reimbursed for costs plus a fee that is adjusted by formula in accordance with the relationship that total allowable costs bear to target cost. At the outset there is negotiated a target cost, a target fee, a minimum and a maximum fee, and the adjustment formula.

6. *Cost-plus-incentive-fee contract (incentive based on performance):* A contract under which a contractor is reimbursed for costs plus an incentive to surpass stated performance targets by providing for increases in the fee to the extent that such targets are surpassed and for decreases to the extent that such targets are not met.[2]

[2]Ibid., pp. 4–6.

In almost all cost-plus-fixed-fee contracts the fee portion is fixed or limited in some way. This means that the contractor knows approximately how much he or she will earn from the contract. However, the owner may not fare so well; in other words, the owner has the greatest risk.

The costs paid by the owner are usually direct costs plus some visible overhead costs, whereas the fee the contractor earns covers overhead and fixed costs as well as profit.

Unit-Price Contracts. This type of contract is one where units can be easily defined and can be costed almost as separate entities. Examples include:

Per mile of road
Per yard of concrete
Per unit of condominium
Per floor of apartment or office space
Per turn-key in a housing tract

Notice how each example has two elements: (1) a measure of the work to be done, and (2) a price for that work. The schedule of unit prices includes fixed and overhead costs and profit.

Variations of unit-price contracts include all the variations cited earlier in fixed-price contracts.

Management Contracts. A management contract is a term associated with a concept of management. Rather than an owner contracting to a general contractor, the owner contracts independently to all parties who will participate in the project. One contract will be between the owner and a "contract manager" (see Exhibit 5-1). This person will act as agent for the owner to see that all contracts entered into by the owner are proper and that their specifications are carried out fully.

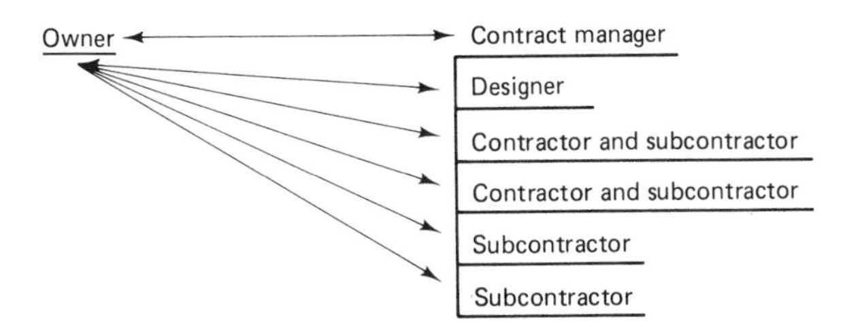

Exhibit 5-1 Contract manager flowchart.

USE OF MANAGEMENT CONTRACTS

1. Very large construction projects
2. Where a large variety of specialized work is needed for a project's completion

3. Where time of the contract is excessively long or phased
4. Where geographical separation makes the use of a general contractor unreasonable

CONTRACT MANAGER RESPONSIBILITIES

1. Organize a staff to maintain cost figures, schedules, and so on, of all contractor and subcontractor contracts.
2. Prepare cost estimates.
3. Prepare construction schedules.
4. Prepare budgets.
5. Provide economic and technical know-how as required.
6. Prepare status or activity reports.
7. Set up conferences between various interested parties.

Government Contracts. The principal difference between two parties entering into a contract and one party entering into the contract with the government is the requirement to meet all of the societal and legal obligations. Since the government must by law apply its own law to itself, it spends vast quantities of time and resources ensuring that the laws are complied with.

This means that a builder, for example, would need to follow *all* of the rules outlined in the numerous segments of the contract to have an opportunity to win the contract. To this end, every facet of a job is covered by forms and instructions.

PARTS OF A GOVERNMENT CONTRACT

1. Invitation for bids (construction contract), form
2. Bid form (construction contract), form
3. Instructions to bidder, form
4. Notice to bidders, form
5. Construction contract, form
6. General provision (construction contract), form
7. Representation and certifications, form
8. Labor standards provisions, form
9. Bid bond
10. Performance bond, form
11. Payment bond, form
12. Amendment to invitation to bid, form

FORMS. The forms illustrated in Figures 5-2 through 5-6 are Postal Service forms, but they illustrate the types of information usually included.

1. Invitation to bid (Figure 5-2)

U.S. POSTAL SERVICE
INVITATION FOR BID
(Construction Contract)

1. INVITATION NO.

2. DATE ISSUED

3. ISSUING OFFICE NAME AND ADDRESS

4. NAME AND LOCATION OF PROJECT

5. THIS INVITATION IS COMPRISED OF THIS FORM AND THE FOLLOWING

	PS Form 7389, Instructions to Bidders		PS Form 7391, General Provisions
	PS Form 7389-A, Notice to Bidders		PS Form 7319-B, Representations and Certifications
	PS Form 7388, Bid Form		PS Form 7322-A, Labor Standards Provisions
	PS Form 7390, Construction Contract		

OTHER *(Specify)*

6. BIDS

Sealed Bids in original and _______ copies for the work described herein will be received at the place specified in Item 3, or if handcarried, in the depository located in ____________________

____________________ until *(Date)* ____________________ 197___, *(Time)* ____________________ local time at the place designated in Item 3 at that time publicly opened. Envelopes containing bids shall be marked to show bidder's name and address, the invitation number and the time and date specified above for receipt of bids. CAUTION: Late bids are subject to the provisions of Paragraph 8, Instructions to Bidders.

7. BIDDING MATERIAL, BID GUARANTEE AND BONDS

8. DESCRIPTION OF WORK AND ESTIMATED COST RANGE *(If in excess of $25,000)*

PS Form 7387
July 1974

Figure 5-2 Invitation to bid. *(continued)*

8. DESCRIPTION OF WORK AND ESTIMATED COST RANGE *(If in excess of $25,000) (Continued)*

NOTICE OF REQUIREMENT FOR CERTIFICATION OF NONSEGREGATED FACILITIES

Bidders and offerors are cautioned as follows:

By signing this bid or offer, the bidder or offeror will be deemed to have signed and agreed to the provisions of the Certification of Nonsegregated Facilities in this solicitation. The certification provides that the bidder or offeror does not maintain or provide for his employees facilities which are segregated on a basis of race, creed, color, or national origin, whether such facilities are segregated by directive or on a de facto basis. The certification also provides that he will not maintain such segregated facilities. Failure of a bidder or offeror to agree to the Certification of Nonsegregated Facilities will render his bid or offer nonresponsive to the terms of solicitations involving awards of contracts exceeding $10,000 which are not exempt from the provisions of the Equal Opportunity clause.

Figure 5-2 *(continued)*

2. Bid form (Figure 5-3)

U.S. POSTAL SERVICE
BID FORM
(Construction Contract)

● THIS FORM TO BE COMPLETED BY BIDDER

1. INVITATION NO.

2. DATE ISSUED

3. BIDDERS NAME AND ADDRESS

4. NAME AND LOCATION OF PROJECT

5. ISSUING OFFICE NAME AND ADDRESS

6. SUBMIT BIDS TO

CAUTION: Bids should not be qualified by exceptions to bidding conditions.

7. BID

DATE SUBMITTED

The undersigned proposes to perform all work included in the above-dated Invitation For Bid in strict compliance with the Invitation, General Provisions (PS Form 7391), specifications, schedules, drawings and conditions for the amount/s shown below:

TOTAL BID: *(In Words)* ______________________________ $ ______________

8. DETAILED PRICES *(Enter detailed prices below, clearly identifying work to which prices apply.*

PS Form 7388
May 1975

Figure 5-3 Bid form. *(continued)*

8. DETAILED PRICES *(Continued)*

9.

The undersigned agrees that, upon written acceptance of this bid, mailed or otherwise furnished within ______ calendar days after the date of opening of bids, he will within ______ calendar days after receipt of the prescribed forms, execute Form 7390, Construction Contract, and give performance and payment bonds on Postal Service forms with good and sufficient surety.

The undersigned agrees, if awarded the contract, that he will commence the work within ________ calendar days after the date of receipt of notice to proceed, and to complete the work within _______ calendar days after the date of receipt of notice to proceed.

10. RECEIPT OF AMENDMENTS

The undersigned acknowledges receipt of the following amendments of the invitation for bids, drawings, and/or specifications, etc. *(Give number and date of each):*

11. BID GUARANTEE

ENCLOSED IS BID GUARANTEE IN THE AMOUNT OF $ ________________ CONSISTING OF: *(Describe)*

12. **The representations and certifications on the accompanying Form 7319-B are made part of this bid.**

13. NAME AND ADDRESS OF BIDDER	14. FULL NAME OF ALL PARTNERS IF BIDDER IS NOT INCORPORATED
15*a*. AUTHORIZED SIGNATURE	15*b*. TYPED NAME AND TITLE

Figure 5-3 *(continued)*

3. Instructions to bidders (Figure 5-4)

U.S. POSTAL SERVICE
INSTRUCTIONS TO BIDDERS
(Construction Contract)

IMPORTANT
READ CAREFULLY

1. Explanations to Bidders. Any explanation desired by a bidder regarding the meaning or interpretation of the invitation for bids, drawings, specifications, etc., must be requested in writing and with sufficient time allowed for a reply to reach bidders before the submission of their bids. Any interpretation made will be in the form of an amendment of the invitation for bids, drawings, specifications, etc., and will be furnished to all prospective bidders. Its receipt by the bidder must be acknowledged in the space provided on the Bid Form (PS Form 7388) or by letter or telegram received before the time set for opening of bids. Oral explanations or instructions given before the award for the contract will not be binding.

2. Conditions Affecting the Work. Bidders should visit the site and take such other steps as may be reasonably necessary to ascertain the nature and location of the work, and the general and local conditions which can affect the work or the cost thereof. Failure to do so will not relieve bidders from responsibility for estimating properly the difficulty or cost of successfully performing the work. The Postal Service will assume no responsibility for any understanding or representations concerning conditions made by any of its officers or agents prior to the execution of the contract, unless included in the invitation for bids, the specifications, or related documents.

3. Bidder's Qualifications. Before a bid is considered for award, the bidder may be requested by the Postal Service to submit a statement regarding his previous experience in performing comparable work, his business and technical organization, financial resources, and plant available to be used in performing the work.

4. Bid Guarantee. Where a bid guarantee is required by the invitation for bids, failure to furnish a bid guarantee in the proper form and amount, by the time set for opening of bids, may be cause for rejection of the bid.

A bid guarantee shall be in the form of a firm commitment, such as a bid bond, postal money order, certified check, cashier's check, irrevocable letter of credit or, in accordance with Treasury Department regulations, certain bonds or notes of the United States. Bid guarantees, other than bid bonds, will be returned (a) to unsuccessful bidders as soon as practicable after the opening of bids, and (b) to the successful bidder upon execution of such further contractual documents and bonds as may be required by the bid as accepted.

If the successful bidder, upon acceptance of his bid by the Postal Service within the period specified therein for acceptance (sixty days if no period is specified) fails to execute such further contractual documents, if any, and give such bond(s) as may be required by the terms of the bid as accepted within the time specified (ten days if no period is specified) after receipt of the forms by him, his contract may be terminated for default. In such event he shall be liable for any cost of procuring the work which exceeds the amount of his bid, and the bid guarantee shall be available toward offsetting such difference. If the contract is not so terminated, the period of delay in providing such bonds may be deducted, at the Contracting Officer's sole discretion, from the time otherwise provided after receipt of the notice to proceed to complete the work.

5. Preparation of Bids. (a) Bids shall be submitted on the forms furnished, or copies thereof, and must be manually signed. If erasures or corrections are made on the forms, each erasure or correction must be initialed by the persons signing the bid. Unless specifically authorized in the invitation for bids, telegraphic bids will not be considered. (b) The bid form may provide for submission of a price or prices for one or more items, which may be lump sum bids, alternate prices, scheduled items resulting in a bid on a unit of construction or a combination thereof, etc. Where the bid form explicitly requires that the bidder bid on all items, failure to do so will disqualify the bid. When submission of a price on all items is not required, bidders should insert the words "no bid" in the space provided for any item on which no price is submitted. (c) Unless called for, alternate bids will not be considered.

6. Submission of Bids. (a) Bids and modifications thereof shall be enclosed in a sealed envelope addressed to the office specified in the solicitations. The bidder shall show the hour and date specified in the solicitation for receipt, the solicitation number, and the name and address of the bidder on the face of the envelope. (b) Telegraphic bids will not be considered unless authorized by the solicitation. Telegraphic modifications should not reveal the amount of the original or the revised bid.

7. Modification or Withdrawal of Bids. Bids already submitted may be modified or withdrawn by written or telegraphic notice, prepared as described in 6. above and received prior to the exact hour and date specified for receipt of bids. A bid may also be withdrawn in person by a bidder or his authorized representative, provided his identity is established and he signs a receipt, but only if the withdrawal is made prior to the exact hour and date set for receipt of bids.

8. Late Offers, Modifications, and Withdrawals. (a) Bids and modifications of bids or withdrawals thereof received at the office designated in the solicitation after the exact hour and date specified for receipt will not be considered unless they are received before award is made and either: (i) they were sent by registered or certified mail not later than the fifth calendar day prior to the date specified for receipt (see paragraph 6 above) or (ii) they were sent by mail (or telegram if authorized) or (iii) they were delivered by other means to the precise depository prescribed in the solicitation and it is determined by the Contracting Officer that the late receipt was due solely to mishandling after receipt by the addressee. (b) The only acceptable evidence to establish– (i) the date of mailing under (a) (i) above is a legible, original postmark supplied and affixed on the date of mailing by Postal Service employees on the bidder wrapper or on the original receipt given therefore, or (ii) the time of receipt under (a) (ii) above is the time/date stamp of such facility on the offer wrapper or other contemporary, documentary evidence of receipt maintained by the facility. (c) Notwithstanding the above, a modification of an otherwise successful bid which makes its terms more favorable to the Postal Service will be considered at any time it is received and may be accepted.

9. Public Opening of Bids. Bids will be publicly opened at the time set for opening in the invitation for bids. Their content will be made public for the information of bidders and others interested, who may be present either in person or by representative.

10. Award of Contract. (a) Award of contract will be made to that responsible bidder whose bid, conforming to the invitation for bids, is most advantageous to the Postal Service, price and other factors considered. (b) The Postal Service may, when in its

PS Form 7389
May 1975

Figure 5-4 Instructions to bidders. (*continued*)

interest, reject any or all bids or waive any informality or minor irregularity in bids received. (c) The Postal Service may accept any item or combination of items of a bid, unless precluded by the invitation for bids or when the bidder includes in his bid a restrictive limitation.

11. Contract and Bonds. The bidder whose bid is accepted will, within the time established in the bid, enter into a written contract with the Postal Service and, if required, furnish performance and payment bonds on Postal Service forms in the amounts indicated in the invitation for bids of the specifications.

OTHER INSTRUCTIONS

Figure 5-4 *(continued)*

4. Construction contract (Figure 5-5)

U.S. POSTAL SERVICE

CONSTRUCTION CONTRACT

● **See Instructions on Reverse**

1. CONTRACT NO.

2. DATE OF CONTRACT

3. CONTRACTOR NAME AND ADDRESS

4. CHECK APPROPRIATE BOX
   - ☐ INDIVIDUAL
   - ☐ PARTNERSHIP
   - ☐ JOINT VENTURE
   - ☐ CORPORATION, INCORPORATED IN THE STATE OF ______

5. ISSUING OFFICE *(Name and Address)*

6. CONTRACT FOR *(Work to be performed)*

7. PLACE

8. CONTRACT PRICE *(Expressed in words and figures)*

9. WORK SHALL BE STARTED *(Date)*

10. WORK SHALL BE COMPLETED *(Date)*

11. FISCAL DATA

12. The United States Postal Service *(hereinafter called the Postal Service)*, represented by the Contracting Officer executing this contract, and the individual, partnership, joint venture, or corporation named above *(hereinafter called the Contractor)*, mutually agree to perform this contract in strict accordance with the General Provisions *(PS Form 7391)*, and the following designated specifications, schedules, drawings, and conditions:

PS Form 7390
July 1974

Figure 5-5 Construction contract. (*continued*)

13. ALTERATIONS. The following alterations were made in this contract before it was signed by the parties hereto:

14. IN WITNESS WHEREOF, THE PARTIES HERETO HAVE EXECUTED THIS CONTRACT AS OF THE DATE ENTERED ON THE FIRST PAGE HEREOF.

THE UNITED STATES POSTAL SERVICE	CONTRACTOR NAME OF CONTRACTOR
SIGNATURE	SIGNATURE BY
TYPED NAME	TYPED NAME
OFFICIAL TITLE	TITLE

INSTRUCTIONS

1. The full name and business address of the Contractor must be inserted in the space provided on the face of the form. The Contractor shall sign in the space provided above with his usual signature and typewrite or print his name under the signature.
2. An officer of a corporation, member of a partnership, or agent signing for the Contractor shall place his signature, typed name and title under the name of the Contractor. A contract executed by an attorney or agent on behalf of the Contractor shall be accompanied by two authenticated copies of his power of attorney or other evidence of his authority to act on behalf of the Contractor.

Figure 5-5 *(continued)*

5. General provisions (index only) (Figure 5-6)

U.S. POSTAL SERVICE

GENERAL PROVISIONS FOR FIXED-PRICE CONSTRUCTION CONTRACTS

INDEX

*Revised

PS Form 7391 Oct. 1979

1

Figure 5-6 General provisions (index only).

Construction Contract Process

1. Responding to an invitation to bid
2. Sending a letter of intent to bid and post bid bond if required
3. Receipt of building/construction project specifications
4. Estimating costs
5. Preparation of bid in prescribed form, and responding in the prescribed manner
6. Award of contract
7. Performance bond obtained
8. Payment bond obtained
9. Performance of contract
10. Modifications/changes (if applicable)
11. Final inspection/acceptance and payment

LEGAL LIABILITY

Today's contractor or builder is subject to the possibility of legal problems from every quarter. Owners can sue for a variety of reasons; workers can get hurt and file claims; pedestrians can be injured at the job site; materials may be defective; and breaches of contracts may take place.

All contractors need to attend one or more courses in business law, contract law, labor relations, and others. Only then can they appreciate their positions. In the next several pages we shall examine very briefly some of the most apparent legal liabilities faced by contractors.

Sole Proprietorship (Unincorporated). The owner of a contracting company, tradesman, and so on, of the sole proprietorship has *unlimited liability*, or legal responsibility for all obligations incurred in doing business.

Partnership (Unincorporated). The partners have *unlimited liability*, just like the sole proprietor.

A *limited partner* in a partnership has liability to the extent of the limited partnership.

Corporation Stockholders. There is limited shareholder liability; however, shareholders are *not* liable for the debts of the corporation.

Note: A corporation may be sued as if it were a person.

Personal Liability. Small companies, regardless of form, may need to agree to sign a note of personal liability reaching into the personal properties of the owners to secure operating capital. This would cover:

1. Homes and furnishings
2. Stocks and negotiable assets
3. Signature value

Negligence Liability. Negligence is defined as omitting to do something that a reasonable person would do.

Negligence liability occurs when the contractor fails to exercise reasonable care in a circumstance (e.g., at a building site) that causes injury to someone. Some obvious causes are:

1. Unprotected excavations
2. Unprotected ground workers working under overhead work
3. Careless handling of materials
4. Careless discharge of fluid materials
5. Exposed electrical wiring
6. Defective or unreliable scaffolding
7. Improper use of construction equipment

Breach of Contract. *Definition:* Whenever a party fails to perform part or all of the duties under a contract, that party is in breach of contract. The party who has not breached the contract is entitled to a remedy by law.

CONTRACTOR BREACH OF CONTRACT. A contractor breaches the contract when he or she fails to comply with the specifications and performance of the contract. Breaches occur by:

1. Using inferior materials
2. Using inferior building methods
3. Not meeting quality construction standards
4. Cheating on quantity of materials
5. Failing to meet time specifications

OWNER BREACH OF CONTRACT. The owner can breach the contract just as easily as a contractor, but may do so because of different reasons. Some are:

1. Failure to make periodic payments
2. Failure to have specific owner responsibilities completed on time
   a. Owner supplied materials late
   b. Owner securing of permits and licenses delayed, causing starting delays and/or idle workers
3. Failure to make known situations affecting the contractor's ability to proceed
4. Before beginning, during, and after performance

TYPES OF REMEDIES. The nonbreached may seek remedy by:

1. Suing for damages to compensate for loss incurred in the endeavor.
   a. Compensatory damages for injuries sustained.
   b. Nominal damages for technical breach of contract.
   c. Consequential or special damages.
2. Initiating *rescission* and making *restitution.*
   a. Rescission action causes an undoing or going back to a prior stage—in effect, the nonbreaching party can rescind the contract.
   b. Restitution, the second half, requires both parties to make restitution to each other or pay back whatever they received.
3. *Specific performance*;
   a. The nonbreached party requires the other party to fulfill the action promised.
   b. This action is advantageous to the nonbreached party.
   c. This action is frequently more desirable than damage remedies in construction contracts.
4. *Reformation* when either party has imperfectly expressed their agreement in writing.
   a. Allows a rewrite of the contract.
   b. Can be used to correct mutual mistakes.
   c. Can be used to overcome fraud in the contract.

Fraud. Any contractor who deliberately misrepresents himself or herself in the preparation or performance of a contract is fraudulently entering the act. The other party is by law allowed to seek civil damages or take criminal action.

Mechanic's Lien. A *mechanic's lien* is placed on real estate by the contractor or subcontractor, when, after the work is completed, the owner is unable to pay. The real estate then becomes security for the lien (debt).

1. The contractor (lien holder) must file a written notice against the particular property.
2. The lien must be filed within a specified time (usually within 60 to 120 days).
3. Failure of the owner to pay allows the lienholder to foreclose on the real estate and sell it to pay off the debt owed.
4. Any monies received in excess of the debt are paid to the former owners.

Default. *Default* occurs, for example, when a contractor who has obtained a builder's note to finance the construction of a project or house fails to make repayment at the appointed time.

1. The lender can and usually does foreclose to recover the loan.
2. The contractor may have no alternative but bankruptcy.
3. A new security agreement may be established.

EXAMPLE: Let us use as an example the building of a house. The builder plans, with the lending institution, to build a house and the institution will advance funds at periodic times during construction and expects the note to be repaid upon the sale of the house not later than 90 days after completion.

Period one: after the house is "dried in."
Period two: after all roughing-in is complete.

Default occurred between periods one and two. The contractor underestimated the materials cost *and* lost $1000 in stolen materials, thus could not meet the specifications for his second loan payment.

1. The institution could refuse the second payment, wait the appropriate amount of time, and then foreclose.
2. The institution could restructure the note and advance the funds.
3. The contractor could try to find a buyer who would advance sufficient down payment to make up the short fall.
4. The contractor could declare bankruptcy.

ACCOUNTS PAYABLE AND NOTES PAYABLE

Both accounts payable and notes payable are forms of indebtness to any contractor. *Accounts payable* are short-term liabilities accepted by contractors. Notes payable are long-term liabilities, usually 1 year or longer.

In the event that a contractor fails to perform and goes into default, these creditors who have extended the materials and monies have early claims for recovery from the contractor.

BONDS

All forms of *bonds* are forms or certificates of a debt on which the issuing company promises to pay the bondholder(s) a specific amount. The payment(s), which may include interest and principal, may be associated with a penalty for nonperformance.

Completion Bond. A completion bond is a form of surety or guarantee agreement which contains the promise of a third party to complete or pay for the cost of completion of a construction contract if the contractor defaults. The contract may be required to obtain this type of bond before the contracts are signed or coincidental with the contract signing. Government construction contracts usually require completion bonding for contracts in excess of $2500.

Contract Bond. A *contract bond* is a guarantee of faithful performance of a construction contract and the payment of all material and labor costs associated with the contract.

PERFORMANCE BOND. If only the performance of a contractor is to be promised, the bond is more properly called a performance bond. This bond ensures that the owner will receive compensation against loss due to the inability or refusal of a contractor to perform.

PAYMENT BOND. If only the material costs and labor costs associated with a contract are to be insured, a payment bond is obtained by the contractor.

Although not frequently done, a contractor may require the owner to post a payment bond with the signing of the contract to ensure his or her payment as the work proceeds or at its completion.

Bid Bond. This is a type of bond that is usually restricted to public construction contracting. When a contractor makes a bid, he or she files the bid bond at the same time. This protects the public in the event that the contractor is awarded the contract and then refuses to enter into the contract. This bond may also apply to withdrawing the bid prior to contract awarding.

Debenture Bond. This is a bond that is secured by the general credit worthiness of a corporation rather than by its specific property.

Mortgage Bond. This is a bond secured by a mortgage on a property.

LEASEHOLD MORTGAGE BOND. This is a bond secured by a building constructed on leased real estate.

Suretyship Bond. This is a three-party bond. The obligator or guarantor agrees to pay a second party upon default by a third party in the performance the third party owes to the second party.

LICENSING OF CONTRACTING COMPANIES

Corporations. To operate a corporation within a state, the owners must file with the secretary of state and obtain legal status. Local permits and/or licenses are often needed.

Partnerships and Sole Proprietorships. Partnerships and sole proprietors must meet state, county, and city or municipality codes to operate as an unincorporated business.

Some types of small contracting companies do not need licensing, others do. Persons planning to engage in these activities need to check with their local courts or licensing bureau before beginning business.

Tax Number. Most states and some local governments issue tax numbers to licensed contractors, which reduce state and local tax liability to the contractor. But it obligates the contractor to collect state and local taxes on any contracts performed within the licensed geographical area.

Employer Identification Number. The employer identification number (EIN) is a requirement by the IRS for:

1. Corporations
2. Partnerships
3. Subchapter S corporations
4. Sole proprietors who
   a. Have employees
   b. Are required to file any excise tax returns

Sole proprietors who do not meet condition 4a or 4b will use their personal social security card as their employee ID number.

To apply for an EIN, simply fill out and mail Form SS4, *Application for Employer Identification Form*, to the IRS.

TAX COMPLIANCE

All contracting companies, regardless of form, are obligated to comply with federal, state, and local tax laws.

Taxable Year. Each construction company needs to establish a taxable year. The company may use:

1. The calendar year
2. A fiscal year (i.e., October 1, 198X to September 30, 198X)
3. A 52-53-week year (see Tax Publication 334)

Federal Filing Requirements for Contractors. Contracting companies may be required to file monthly reports with the IRS. They will need to file quarterly and annually.

Whether or not a company earns a profit, the owners or their legal representatives are required to file income tax forms. Checklist 5-1 lists the information and forms required for filing.

Some of the federal taxes for which a sole proprietor, a corporation, or a partnership may be liable are listed below. If a due date falls on a Saturday, Sunday, or legal holiday, it is postponed until the next day that is not a Saturday, Sunday, or legal holiday. For more information, see Publication 509, *Tax Calendars for 1983.*

You may be liable for	If you are:	Use Form	Due on or before
Income tax	Sole proprietor	Schedule C (Form 1040)	Same day as Form 1040
	Individual who is a partner or Subchapter S corporation shareholder	1040	15th day of 4th month after end of tax year
	Corporation	1120	15th day of 3rd month after end of tax year
	Subchapter S corporation	1120S	15th day of 3rd month after end of tax year
Self-employment tax	Sole proprietor, or individual who is a partner	Schedule SE (Form 1040)	Same day as Form 1040
Estimated tax	Sole proprietor, or individual who is a partner or Subchapter S corporation shareholder	1040–ES	15th day of 4th, 6th, and 9th months of tax year, and 15th day of 1st month after the end of tax year
	Corporation	1120W	15th day of 4th, 6th, 9th and 12th months of tax year
Annual return of income	Partnership	1065	15th day of 4th month after end of tax year
FICA tax and the withholding of income tax	Sole proprietor, corporation, Subchapter S corporation, or partnership	941 501 (to make deposits)	4–30, 7–31, 10–31 and 1–31 See Chapter 19
Providing information on FICA tax and the withholding of income tax	Sole proprietor, corporation, Subchapter S corporation, or partnership	W–2 (to employee) W–3 (to the Social Security Administration)	1–31 Last day of February
FUTA tax	Sole proprietor, corporation, Subchapter S corporation, or partnership	940 508 (to make deposits)	1–31 4–30, 7–31, 10–31 and 1–31, but only if the liability for unpaid tax is more than $100
Annual information returns	Sole proprietor, corporation, Subchapter S corporation, or partnership	See Publication 15	
Excise taxes	Sole proprietor, corporation, Subchapter S corporation, or partnership	See Chapter 36	

Checklist 5-1 IRS checklist for tax filing (for all types of contractors). (*Source*: IRS Publication 334, Tax Guide for Small Businesses)

TAX PUBLICATIONS AND FORMS

The listing of tax publications and forms (shown in Checklist 5-2) may be of use. Each of these is easily obtained from local IRS offices by simply phoning and asking.

Tax Publications

On the preceding page, there is an order blank that you can use to send for any other IRS tax publications you need. Many of the publications that might be of interest to business taxpayers are listed below by their numbers and titles. This list includes all the publications referred to in the chapters of this book. A full list can be found in Publication 910, *Taxpayer's Guide to IRS Information and Assistance.*

General Guides:

17 .... Your Federal Income Tax (an income tax guide for individuals)
225 .... Farmer's Tax Guide
509 .... Tax Calendars for 1983
553 .... Highlights of 1982 Tax Changes
595 .... Tax Guide for Commercial Fishermen
910 .... Taxpayer's Guide to IRS Information and Assistance

Employer's Guides:

15 .... Employer's Tax Guide (Circular E)
51 .... Agricultural Employer's Tax Guide (Circular A)
80 .... Federal Tax Guide for Employers in the Virgin Islands, Guam, and American Samoa (Circular SS)

Specialized Publications:

349 .... Federal Highway Use Tax on Trucks, Truck Tractors, and Buses
378 .... Fuel Tax Credits
463 .... Travel, Entertainment, and Gift Expenses
505 .... Tax Withholding and Estimated Tax
506 .... Income Averaging
510 .... Excise Taxes for 1983
517 .... Social Security for Members of the Clergy and Religious Workers
521 .... Moving Expenses
523 .... Tax Information on Selling Your Home
525 .... Taxable and Nontaxable Income
526 .... Charitable Contributions
527 .... Rental Property
529 .... Miscellaneous Deductions
531 .... Reporting Income from Tips
533 .... Self-Employment Tax
534 .... Depreciation
535 .... Business Expenses
536 .... Net Operating Losses and the At-Risk Limits
537 .... Installment Sales
538 .... Accounting Periods and Methods
539 .... Employment Taxes—Income Tax Withholding, FICA and FUTA, Advance Payments of EIC, Withholding on Gambling Winnings
541 .... Tax Information on Partnerships
542 .... Tax Information on Corporations
544 .... Sales and Other Dispositions of Assets
545 .... Interest Expense
547 .... Tax Information on Disasters, Casualties, and Thefts
548 .... Deduction for Bad Debts
549 .... Condemnations of Private Property for Public Use
550 .... Investment Income and Expenses
551 .... Basis of Assets
556 .... Examination of Returns, Appeal Rights, and Claims for Refund
560 .... Tax Information on Self-Employed Retirement Plans
561 .... Determining the Value of Donated Property
572 .... Investment Credit
575 .... Pension and Annuity Income
583 .... Information for Business Taxpayers—Business Taxes, Identification Numbers, Recordkeeping
584 .... Disaster and Casualty Loss Workbook
586A .... The Collection Process (Income Tax Accounts)
587 .... Business Use of Your Home
589 .... Tax Information on Subchapter S Corporations
590 .... Tax Information on Individual Retirement Arrangements
597 .... Information on the United States-Canada Income Tax Treaty
598 .... Tax on Unrelated Business Income of Exempt Organizations
906 .... Jobs and Research Credits
908 .... Bankruptcy
909 .... Minimum Tax and Alternative Minimum Tax
911 .... Tax Information for Direct Sellers
1045 .... Information and Order Blanks for Preparers of Federal Income Tax Returns
1048 .... Filing Requirements for Employee Benefit Plans

Checklist 5-2 IRS publications and forms listing. (*Source*: IRS Publication 334, Tax Guide for Small Businesses)

The most important forms for the various types of contracting companies are:

Sole Proprietorship

1. Schedule C (Form 1040)
2. Form 1040 (basic)
3. Form 4562, depreciation and amortization
4. Form 3468, computation of investment credit

Partnership

1. Form 1065, partnership (with all schedules)
2. Form 1040, each partner
3. Form 4797, supplemental schedule of gains and losses
4. Form 3468
5. Form 4562

Corporation

1. Form 1120, corporation
2. Form 4797, supplemental schedule of gains and losses
3. Form 4562, depreciation and amortization
4. Form 4466, corporation application for quick refund of overpayment of estimated tax
5. Form 7004, application for automatic extension of time to file corporation income tax form
6. Form 7005, same title as 7004
7. Form 2439, notice to shareholder of undistributed long-term capital gain
8. Form 4136, computation for federal tax on gasoline, special fuels, and lubricating oil
9. Form 2220, underpayment of estimated tax by corporation
10. Form 5471
11. Form 1118, foreign tax credit
12. Form 3468, computation of investment credit
13. Form 5884, job credits
14. Form 6765, research credits
15. Form 5735, possessions tax credits
16. Form 6478, alcohol fuel credits
17. Form 4255, prior-year investment credit
18. Form 4626, minimum tax on tax preference items
19. Form 5452, corporate report of nontaxable dividends
20. Form 851, affiliations schedule

Subchapter S Corporation

1. Form 1120S, Subchapter S
2. Form 4797
3. Form 4562
4. Forms 7004, 7005
5. Form 4136
6. Forms 3520, A, or 926:
   a. 3520, creation of or transfers to certain foreign trusts
   b. 926, return by a transfer to property to a foreign corporation, foreign trust, or foreign partnership.
7. Form 3468

BUSINESS TAXES OTHER THAN FEDERAL INCOME TAX

Those contracting firms that employ workers, office help, and technical specialists or managers must participate in federal and state taxing programs that benefit these employees. Those who are sole proprietors may also want to insure themselves for old age and Medicare benefits.

Self-Employment Tax. A sole-proprietor or partner contractor may have to file Schedule SE (of Form 1040) together with Form 1040. This is a social security tax for contractors who work for themselves. This tax is a pay-as-you-go type, which means that the contractor pays into the account as earnings are incurred. If the self-employment tax is expected to exceed $300 per annum, estimated quarterly payments may be required (see IRS Publications 533 and 505).

Employment Taxes for All forms of Contracting Companies

1. Federal and state income tax withheld from employees wages (see IRS Publication 15, covering FITW and SITW).
2. Social security (FICA) tax includes *both* the amount withheld from employees' wages and the contractor's (employer's) contribution to the employees' accounts.
3. Federal *and* state unemployment tax. Federal (FUTA) and state unemployment tax is money paid to the federal and state governments based on the wages paid employees. These monies are held separate from FICA and FITW and state withholdings. As a rule, an account is set up with a base amount. As monthly payments to employees become due, payments are made from the account; then the employer replenishes the account.

The account's base is computed differently from state to state. The source of information is the State Employment Security Commission for each state. The account's base requires adjustment as the number of employees changes.

Note: Beginning in January 1983, employers pay FUTA at a rate of 3.5%, as well as state unemployment compensation according to state formulas.

REFERENCES AND SOURCES

1. Educational sources
   a. Institutions
      (1) Junior and senior colleges
      (2) Graduate programs
   b. Subjects
      (1) Business I and II
      (2) Advanced business law
      (3) Money—banking and policy
      (4) Accounting and financial reporting
      (5) Tax and tax structure
      (6) Business finance
      (7) Business policy and administration

2. Other sources
   a. American Institute of Certified Public Accountants, Inc. (AICPA), 1211 Avenue of the Americas, New York, NY 10036
   b. Tax guides for business (IRS)
   c. Public and law libraries
   d. Prentice-Hall, Inc., Englewood Cliffs, NJ 07632
   e. McGraw-Hill Book Company, 1221 Avenue of the Americas, New York, NY 10020

6

INSURANCE, SECURITY, AND CONSTRUCTION SAFETY

QUICK REFERENCES

This chapter provides information on ways in which a contractor or contracting company can protect its interests. In every circumstance, insurance, security, and safety represent a trade-off: controlled expenses versus possible catastrophic expenses or failure of the company. Recall from Chapter 2 the definitions of risk and uncertainty. Insurance security and safety practices affect the company's risk and lessen its uncertainty. The trade-off involves how much risk a contractor should assume versus how much of the three subjects of this chapter he or she should buy with company dollars.

INSURANCE

One of the many costs of operating a construction company is the insurance expense. In this section of this chapter, we describe briefly the variety of insurance needs and types of insurance that are most beneficial to the contractor, the workers, and his or her investments. A rule of thumb is to buy as much insurance as is necessary to cover the risk or uncertainty at the lowest possible price. To do this, each contractor must do two things: (1) he or she must carefully define the amount of coverage needed for each type of insurance; (2) he or she needs to eliminate as much as possible the risk or uncertainty associated with the purposes of insurance, thereby taking maximum advantage of the lowest premiums available.

Definition

Insurance is a contract with the insurance company (insurer) that promises to pay a sum of money or to protect, repair, or make good the owners (insured) property in the event that the insured sustains damage or injury as covered by the insurance policy.

In another way, insurance is a contract that the insured buys to lower or minimize a potential loss situation. If, for example, statistically, a non-employee at the job site is injured in 1 of 10 jobs, a contractor would assume that his or her risk is 1 in 10 or 10%. If the severity of the possible injury is expected to be minimal, a small public liability coverage policy would be adequate. However, if the values stated above were significantly higher, a very large policy would be needed and would be a costly expense to the contract.

Terms

Binder. The insurance agent or broker will generally write a memorandum or *binder* indicating that a contract is in force, a policy is forthcoming, and the essential terms of the contract.

Builder's Risk. A *builder's risk* policy is a general umbrella policy encompassing all forms of risk that a builder may anticipate. Recall that *risk* is the statistical probability that an event may or may not occur, whereas *uncertainty* is the best guess available.

Indemnity. *Indemnity* is insurance against possible loss, damage, or injury. A contractor would indemnify himself or herself from possible damage, law suits, or bodily injury by taking out an insurance (indemnity) policy. As a general rule, fire insurance policies are written as indemnity policies.

Insurable Interest. A contractor has an *insurable interest* in property if damage to or destruction of the property will cause direct monetary loss. For example, a contractor insures against building materials loss at the job site. Since he or she owns the materials (even if they are still on accounts payable), there is an insurable interest.

Insured and Insurer. The contractor is usually the *insured* and the insurance company is the *insurer*.

Insurance Broker. An insurance *broker* is usually an independent contractor who sells a variety of insurance. He or she usually employ *agents*.

Public Liability. All contractors as a rule are insured for *public liability*. This covers third parties—nonemployees or pedestrians who may become injured on the job or in connection with the contracting company's property or personnel.

Surety. The general definition of surety is a pledge made to secure against loss, damage, or default; a form of guarantee. With regard to contractors, *surety* is more properly termed "suretyship," which is a person or company who has contracted to be responsible for another's (say ABC Contractor's) obligations or debts in the event of the default of ABC Contractors.

Suretyship may also apply to the owners who are contracting the work done. Both types of surety are illustrated below.

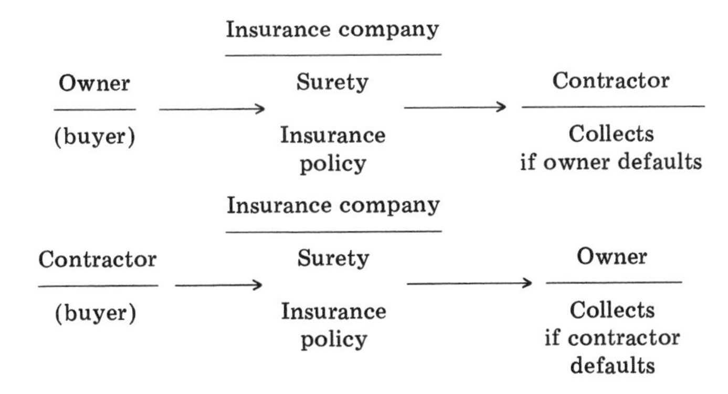

Warranty. A warranty is a form of insurance that a contractor may elect to provide or be required to provide. In essence, it promises that the building or other project shall conform to specifications and that title to such property belongs to the buyer when all parts of the contract are fulfilled. Various types of warranties include infringement, expressed, implied, exclusionary, and third party.

Classification or Types of Insurance

In some respects the classification of insurance available to a contractor is difficult to catalog. This is primarily because so many policies are issued as "umbrella" coverage or are termed "package plans." Nevertheless, if a contractor determines that he or she has a risk of uncertainty that needs protection, personal property, fire, and casualty insurance may be the answer. We shall briefly identify the variety of insurance available. Many may be attached as *clauses* or *riders* to other basic policies

Business. Business insurance coverage is obtained to lessen the risk associated with operations.

ACCOUNTS RECEIVABLE. To insure accounts receivable, one assumes that there is a particular risk that some creditors may not pay their accounts and/or that allowances for doubtful accounts would be too costly to the business.

BUSINESS INTERRUPTION. This type of policy may be a contingency type that indemnifies the contractor for loss of gross earnings and other such charges and expenses resulting from the necessary interruption of the business. The interruption may be due to damage or destruction, such as fire or casualty. If the policy is a contingency type, an insured peril at the business or job site must occur if the policy is to take force. This policy may be attached to a fire protection policy. Raw materials, work in process, and even ordinary payroll expenses may be covered.

BUILDINGS AND CONTENTS. This policy would probably be a blanket or umbrella type. It could cover office and office furniture and fixtures, or it could cover a warehouse and stored inventory; or it could cover a shop, its machines and inventory.

BUSINESS OWNER'S POLICY. An umbrella policy that provides a wide range of coverage. The advantage: broad coverage at values the insurer selects. The disadvantage: coverage may not be adequate to the contractor's risk or may be more than needed, thereby increasing costs on the latter.

CONTINGENT LIABILITY FROM OPERATIONS OF BUILDING LAWS. This type of insurance provides for occurrences where damaged property must be reconfigured to present building codes (i.e., adding automatic fire sprinklers, ramps for disabled persons, or conformance to new electrical codes).

CONTRACTOR EQUIPMENT COVERAGE. Leased or purchased equipment can be insured for various coverages: for example, a specific peril or an all-risk policy. Most frequently these policies are issued for heavy equipment; however, almost any on-site equipment can be insured. Insurers may suggest that such coverage be attached to a basic policy.

EARNINGS. This policy is a limited version of business interruption coverage. It requires less reporting but pays lower percentages of earnings losses.

FLEET. Any contracting firm that has a fleet of trucks may purchase fleet insurance. This type of insurance allows insurers to provide discounts based on numbers and types of vehicles. Naturally, a wide variety of coverage is available.

FLOOD. Insurance coverage associated with a risk of damage by rising waters is called flood insurance. As a rule, this is expensive insurance and is not always available. Contractors should be wary if they are building in flood-prone areas, at least during the construction phase. They may also want to avoid building in low-laying areas where flood coverage is not available.

Fire Indemnity Coverage. This policy is really a basic policy which most contractors obtain. It can be designed for specified items, such as materials, buildings, furniture and fixtures, or remote-site locations. But more than likely, a blanket policy would be cheaper. Then attachments, riders, or floaters could be added or deleted as needs changed.

Increased Cost of Construction. This type of insurance is very similar to contingent liability, described earlier. Purchase of this coverage is often dictated by costs above replacement-value costs necessitated by enforcement of new building codes.

Liability Coverage. Liability coverage is insuring against the possibility of third parties suing for injuries or losses sustained on the job site or business premises. However, the insurer frequently attempts to settle such losses by paying the costs of injury or loss.

Loss of Income. This policy protects the contractor against loss of income if his or her business is destroyed. This coverage is usually made a part of a fire indemnity or business interruption policy.

Payroll Coverage. This coverage is also part of business interruption insurance.

Comprehensive Coverage. This is a blanket policy that covers a multitude of possible losses. Comprehensive coverage may be a part of auto coverage, resident coverage or business property or equipment coverages. Some elements covered might include theft, glass breakage, vandalism, and storm damage.

Casualty Insurance. This type of coverage is a broad category. A contractor may find an all-risk casualty policy for specific items.

Personal Property. This type of insurance is more properly called property damage and bodily injury liability. This is a basic coverage and usually inexpensive, based on per-thousand-dollar coverage. Face values of $300,000 and up are common today. This coverage protects the contractor and his or her business from third-party law suits.

Office Personal Property. This is a liability policy that insures office equipment, desks, and files. Their contents are secured against either all risks or defined limited risks. Usually, the insurer requires a deductible (say $50 or $100) at a premium discount.

Unemployment. The unemployment insurance described here is more fully described in Chapter 5. Suffice it to say that today such insurance is mandatory by the federal government and in most states. The amounts vary from state to state based on a variety of factors. A check with the local employment security commission will answer all questions. The cost of this insurance drives up the cost of doing business.

Worker's Compensation. Worker's compensation insurance covers accident, health, and disability of workers who suffer losses while employed. The insurance is mandatory in all states. It usually covers losses due to physical injury. Recently, diseases have been included in the coverage.

The amount of compensation that an injured or ill worker is paid is determined by a fixed schedule of payment benefits. The factors determining the amount are (1) the wages being earned by the worker, (2) the seriousness of the injury or illness, and (3) the permanence of the injury or illness.

Group Insurance. Contracting firms that employ a large standing crew and office staff may be able to provide group insurance for their employees. Numerous varieties are available. Several of the more frequently selected types are health and accident, and salary continuation.

Premiums paid for coverage of the employees is usually based on the number of employees and the extent of and variety of coverage.

Many companies share the cost of these policies. Employees contribute a small portion and the contractor contributes a share.

Accounting for Insurance Premiums

All insurance policies are prepaid. Therefore, the unused portion of these premiums is an asset that shows up on the balance sheet. However, the expended portion of the premium is an expense of doing business. Because there are so many possible types of policies and coverages, an insurance register or ledger is established. All policies are cataloged in these books so that at any time, the contractor or manager may value his or her accounts. This means that in any given month, the contractor can determine the prepaid value remaining on his or her insurance policy.

Each insurance policy and its riders or attachments should be charged to either general expense, office expense, a specific job, or other indirect expense.

SECURITY

Today's contractor faces a real threat of business collapse if job-site and business-site security is taken for granted. Theft of materials and equipment is a growing problem regardless of the state of the economy. The larger and more concentrated a population is, the greater the problem of security.

We have already alluded to security in one form—insurance. We should mention further that the premiums paid by contractors to insure against theft are based on a contractor's past experiences.

Two particular methods reduce the possibility of loss: a security guard and a secured building.

Security Guard

If job sites are large enough so that hundreds of thousands of dollars in equipment and materials are stored there, a patroling guard or guards are usually hired from a local security service to protect the site. Sometimes, off-duty police or retired law enforcement officers are available.

If the job site is secured by a fence, this may be enough to protect the job. However, guard dogs are frequently allowed to roam freely within the site.

Business properties—general offices—should also be secured by alarms and bars. Alarms are available that are connected electrically to local police stations. These may be silent alarms, audible alarms, or a combination.

Security Expenses

All security measures are expenses of doing business. They should be charged to job sites where applicable, to office expenses if appropriate, or to general expenses.

OCCUPATIONAL SAFETY AND HEALTH ADMINISTRATION

The portion of Occupational Safety and Health Administration (OSHA) Standards that apply to the construction industry is a subsection of *General Industry*, Parts 1910 and 1926. Each contractor would be more knowledgeable if he or she had a copy of these parts (and others that apply to special construction conditions). The address for obtaining a copy is:

Occupational Safety and Health Administration
U.S. Department of Labor
200 Constitution Avenue NW
Washington, DC 20210

Purpose

Most of the following material is reprinted from the Occupational Safety and Health Act of 1970 that became law on December 29, 1970. Its purpose is:

> To assure safe and healthful working conditions for working men and women; by authorizing enforcement of the standards developed under the Act; by assisting and encouraging the States in their efforts to assure safe and healthful working conditions; by providing for research, information, education, and training in the field of occupational safety and health; and for other purposes.
>
> This volume of Occupational Safety and Health Standards and Interpretations is comprised of national consensus standards and established Federal standards. These standards are applicable to the construction industry.
>
> These Construction Industry Standards were originally published in the Federal Register. They were subsequently codified and republished in the Code of Federal Regulations.
>
> Only the Federal Register can be used as the official source or reference for such material. Every effort has been made to repeat verbatim the information published in the Federal Register—however, in the event of conflict or inconsistency, the Federal Register takes precedence as the official publication.
>
> The purposes of publishing this set of standards and interpretations are (1) to facilitate their use through a different and easier format of the material (finding and reading specific sections and paragraphs has been greatly enhanced); and (2) to create a publication system for occupational safety and health standards, interpretations, regulations, and procedures which can be more easily maintained and revised. These standards and interpretations are available through the subscription services of the Superintendent of Documents, U.S. Government Printing Office.

Basic Manual

The basic manual consists of front matter, the text of the standards, and appendices.

1. *Front matter:* The front matter includes the title page, List of Effective Pages, Record of Changes/Additions page, contents, and introduction.
   a. Title page identifies the text of the book, date issued, date of the current change and the volume number.

b. List of Effective Pages or "A" page indicates total number of pages in this volume and lists all pages with change status and page order.
c. Record of Changes/Additions page specifies the number of change, date of change/addition, date entered and signature of person entering change.
d. Table of Contents lists the text of the manual by part and subpart numbers. It also indicates page number for quick reference.
e. Introduction describes the format and contents of this manual.

2. *Text:* The text of the manual includes all of the standards associated with the construction industry. Please note, however, that the construction industry is also subject to the regulations contained in 29 CFR part 1910, general industry standards which are included in Volume I of this service. The standards are divided by part and subpart; each subpart is preceded by a list of sections contained therein. Parts and subparts covered in this manual are as follows: 29 CFR 1926, Safety and Health Regulations for Construction, subparts A thru X.

   The format of the text is changed from that of the Federal Register to enhance reader useability.
3. *Appendices*

Part 1910.XXX: General Industry Standards

Part 1910 applies to the construction industry because many of its subparts logically cover safety and hazards common to the construction site and the workers.

Part 1910 consists of Subparts A through Z with several subparts reserved and others not used. Each subpart is indexed on a lead sheet which lists the sections within the subpart. A listing of Part 1910 follows:

Subpart A—General

Sec.
1910.1 Purpose and scope.
1910.2 Definitions.
1910.3 Petitions for the issuance, amendment, or repeal of a standard.
1910.4 Amendments to this part.
1910.5 Applicability of standards.
1910.6 Incorporation by reference.

Subpart B—Adoption and Extension of Established Federal Standards

1910.11 Scope and purpose.
1910.12 Construction work.
1910.13 Ship repairing.
1910.14 Shipbuilding.
1910.15 Shipbreaking.
1910.16 Longshoring.
1910.17 Effective dates.
1910.18 Changes in established Federal standards.
1910.19 Asbestos dust.

Subpart C—[Reserved]

Subpart D—Walking-Working Surfaces

1910.21 Definitions.
1910.22 General requirements.
1910.23 Guarding floor and wall openings and holes.
1910.24 Fixed industrial stairs.
1910.25 Portable wood ladders.
1910.26 Portable metal ladders.
1910.27 Fixed ladders.
1910.28 Safety requirements for scaffolding.

1910.29 Manually propelled mobile ladder stands and scaffolds (towers).
1910.30 Other working surfaces.
1910.31 Sources of standards.
1910.32 Standards organizations.

Subpart E—Means of Egress

1910.35 Definitions.
1910.36 General requirements.
1910.37 Means of egress, general.
1910.38 Specific means of egress requirements by occupancy. [Reserved]
1910.39 Sources of standards.
1910.40 Standards organizations.

Subpart F—Powered Platforms, Manlifts, and Vehicle-Mounted Work Platforms

1910.68 Power platforms for exterior building maintenance.
1910.67 Vehicle-mounted elevating and rotating work platforms.
1910.68 Manlifts.
1910.69 Sources of standards.
1910.70 Standards organizations.

Subpart G—Occupational Health and Environmental Control

1910.94 Ventilation.
1910.95 Occupational noise exposure.
1910.96 Ionizing radiation.
1910.97 Nonionizing radiation.
1910.98 Effective dates.
1910.99 Sources of standards.
1910.100 Standards organizations.

Subpart H—Hazardous Materials

1910.101 Compressed gases (general requirements).
1910.102 Acetylene.
1910.103 Hydrogen.
1910.104 Oxygen.
1910.105 Nitrous oxide.
1910.106 Flammable and combustible liquids.
1910.107 Spray finishing using flammable and combustible materials.
1910.108 Dip tanks containing flammable or combustible liquids.
1910.109 Explosives and blasting agents.
1910.110 Storage and handling of liquified petroleum gases.
1910.111 Storage and handling of anhydrous ammonia.
1910.112 [Reserved]
1910.113 [Reserved]
1910.114 Effective dates.
1910.115 Sources of standards.
1910.116 Standards organizations.

Subpart I—Personal Protective Equipment

1910.132 General requirements.
1910.133 Eye and face protection.
1910.134 Respiratory protection.
1910.135 Occupational head protection.
1910.136 Occupational foot protection.
1910.137 Electrical protective devices.
1910.138 Effective dates.
1910.139 Sources of standards.
1910.140 Standards organizations.

Subpart J—General Environmental Controls

1910.141 Sanitation.
1910.142 Temporary labor camps.
1910.143 Nonwater carriage disposal systems.
1910.144 Safety color code for marking physical hazards.
1910.145 Specifications for accident prevention signs and tags.
1910.146 [Reserved]
1910.147 Sources of standards.
1910.148 Standards organizations.
1910.149 Effective dates.

Subpart K—Medical and First Aid

1910.151 Medical services and first aid.
1910.152 [Reserved]
1910.153 Sources of standards.

Subpart L—Fire Protection

1910.156 Definitions applicable to this subpart.

Portable Fire Suppression Equipment

1910.157 Portable fire extinguishers.
1910.158 Standpipe and hose systems.

Fixed Fire Suppression Equipment

1910.159 Automatic sprinkler systems.
1910.160 Fixed dry chemical extinguishing systems.
1910.161 Carbon dioxide extinguishing systems.
1910.162 Other special fixed extinguishing systems. [Reserved]

Other Fire Protection Systems

1910.163 Local fire alarm signaling systems.
1910.164 Fire brigades. [Reserved]
1910.165 Effective dates.
1910.165a Sources of standards.
1910.165b Standards organizations.

Subpart M—Compressed Gas and Compressed Air Equipment

1910.166 Inspection of compressed gas cylinders.
1910.167 Safety relief devices for compressed gas cylinders.
1910.168 Safety relief devices for cargo and portable tanks storing compressed gases.
1910.169 Air receivers.
1910.170 Sources of standards.
1910.171 Standards organizations.

Subpart N—Materials Handling and Storage

1910.176 Handling materials—general.
1910.177 [Reserved]
1910.178 Powered industrial trucks.
1910.179 Overhead and gantry cranes.
1910.180 Crawler, locomotive, and truck cranes.
1910.181 Derricks.
1910.182 Effective dates.
1910.183 Helicopters.
1910.184 Slings.
1910.189 Sources of standards.
1910.190 Standards organizations.

Subpart O—Machinery and Machine Guarding

1910.211 Definitions.
1910.212 General requirements for all machines.
1910.213 Woodworking machinery requirements.
1910.214 Cooperage machinery.
1910.215 Abrasive wheel machinery.
1910.216 Mills and calenders in the rubber and plastics industries.
1910.217 Mechanical power presses.
1910.218 Forging machines.
1910.219 Mechanical power-transmission apparatus.
1910.220 Effective dates.
1910.221 Sources of standards.
1910.222 Standards organizations.

Subpart P—Hand and Portable Powered Tools and Other Hand-Held Equipment

1910.241 Definitions.
1910.242 Hand and portable powered tools and equipment, general.
1910.243 Guarding of portable powered tools.
1910.244 Other portable tools and equipment.
1910.245 Effective dates.
1910.246 Sources of standards.
1910.247 Standards organizations.

Subpart Q—Welding, Cutting, and Brazing

1910.251 Definitions.
1910.252 Welding, cutting, and brazing.
1910.253 Sources of standards.
1910.254 Standards organizations.

Subpart R—Special Industries

1910.261 Pulp, paper, and paperboard mills.
1910.262 Textiles.
1910.263 Bakery equipment.
1910.264 Laundry machinery and operations.
1910.265 Sawmills.
1910.266 Pulpwood logging.
1910.267 Agricultural operations.
1910.268 Sources of standards.
1910.269 Standards organizations.
1910.274 Sources of standards.
1910.275 Standards organizations.

Subpart S—Electrical

1910.308 Application.
1910.309 National Electrical Code.

Subpart Z—Toxic and Hazardous Substances

1910.1000 Air Contaminants.
1910.1001 Asbestos.
1910.1002 Coal tar pitch volatiles; interpretation of term.
1910.1003 4-Nitrobiphenyl.
1910.1004 alpha-Naphthylamine.
1910.1005 4,4′—Methylene bis (2-chloroaniline).
1910.1006 Methyl chloromethyl ether.
1910.1007 3,3′—Dichlorobenzidine (and its salts).
1910.1008 bis-Chloromethyl ether.
1910.1009 beta-Naphthylamine.
1910.1010 Benzidine.
1910.1011 4-Aminodiphenyl.
1910.1012 Ethyleneimine.
1910.1013 beta-Propioiactone.
1910.1014 2-Acetylaminofluorene.
1910.1015 4-Dimethylaminoazobenzene.
1910.1016 N-Nitrosodimethylamine.
1910.1017 Vinyl chloride.

Part 1926.XXX

Part 1926, titled *Safety and Health Regulations for Construction* follows. It consists of Subparts A through X. A review of its contents reveals the mixing of Part 1910 with Part 1926.

Subpart A—General

1926.1 Purpose and scope.
1926.2 Variances from safety and health standards.
1926.3 Inspections—right of entry.
1926.4 Rules of practice for administrative adjudications for enforcement of safety and health standards.

Subpart B—General Interpretations

1926.10 Scope of subpart.
1926.11 Coverage under section 103 of the act distinguished.
1926.12 Reorganization Plan No. 14 of 1950.
1926.13 Interpretations of statutory terms.
1926.14 Federal contracts for "mixed" types of performance.
1926.15 Relationship to the Service Contract Act; Walsh-Healey Public Contracts Act.
1926.16 Rules of construction.
1910.11 Scope and purpose.
1910.12 Construction work.
1910.16 Longshoring.
1910.19 Special provisions for air contaminants.

Subpart C—General Safety and Health Provisions

1926.20 General safety and health provisions.
1926.21 Safety training and education.
1926.22 Recording and reporting of injuries. [Reserved]
1926.23 First aid and medical attention.
1926.24 Fire protection and prevention.
1926.25 Housekeeping.
1926.26 Illumination.
1926.27 Sanitation.
1926.28 Personal protective equipment.
1910.132 General requirements (personal protective equipment).
1910.136 Occupational foot protection.
1926.29 Acceptable certifications.
1926.30 Shipbuilding and ship repairing.
1926.31 Incorporation by reference.
1926.32 Definitions.

Subpart D—Occupational Health and Environmental Controls

1926.50 Medical services and first aid.
1926.51 Sanitation.
1910.141 Sanitation.
1910.151 Medical services and first aid.

1926.52 Occupational noise exposure.
1926.53 Ionizing radiation.
1926.54 Nonionizing radiation.
1926.55 Gases, vapors, fumes, dusts, and mists.
1910.161 Carbon dioxide extinguishing systems.
1926.56 Illumination.
1926.57 Ventilation.

Subpart E—Personal Protective and Life Saving Equipment

1926.100 Head protection.
1926.101 Hearing protection.
1926.102 Eye and face protection.
1926.103 Respiratory protection.
1910.94 Ventilation.
1910.134 Respiratory protection.
1926.104 Safety belts, lifelines, and lanyards.
1926.105 Safety nets.
1926.106 Working over or near water.
1926.107 Definitions applicable to this subpart.

Subpart F—Fire Protection and Prevention

1926.150 Fire protection.
1926.151 Fire prevention.
1926.152 Flammable and combustible liquids.
1910.106 Flammable and combustible liquids.
1926.153 Liquefied petroleum gas (LP-Gas).
1910.110 Storage and handling of liquefied petroleum gases.
1926.154 Temporary heating devices.
1926.155 Definitions applicable to this subpart.

Subpart G—Signs, Signals, and Barricades

1926.200 Accident prevention signs and tags.
1926.201 Signaling.
1926.202 Barricades.
1926.203 Definitions applicable to this subpart.

Subpart H—Materials Handling, Storage, Use, and Disposal

1926.250 General requirements for storage.
1910.30 Other working surfaces.
1910.176 Handling materials—general.
1926.251 Rigging equipment for material handling.
1910.184 Slings.
1926.252 Disposal of waste materials.

Subpart I—Tools—Hand and Power

1926.300 General requirements.
1910.212 General requirements for all machines.
1926.301 Hand tools.
1926.302 Power operated hand tools.
1910.244 Other portable tools and equipment (abrasive blast cleaning nozzles).
1926.303 Abrasive wheels and tools.
1926.304 Woodworking tools.
1926.305 Jacks—lever and ratchet, screw and hydraulic.
1910.244 Other portable tools and equipment (jacks).

Subpart J—Welding and Cutting

1926.350 Gas welding and cutting.
1926.351 Arc welding and cutting.
1926.352 Fire prevention.
1926.353 Ventilation and protection in welding, cutting, and heating.
1926.354 Welding, cutting and heating in way of preservative coatings.

Subpart K—Electrical

1926.400 General requirements.
1926.401 Grounding and bonding.
1926.402 Equipment installation and maintenance.
1926.403 Battery rooms and battery charging.
1926.404 Hazardous locations.
1926.405 Definitions applicable to this subpart.

Subpart L—Ladders and Scaffolding

1926.450 Ladders.
1926.451 Scaffolding.
1910.21 Definitions.
1910.28 Safety requirements for scaffolding.

1910.29 Manually propelled mobile ladder stands and scaffolds (towers).
1926.452 Definitions applicable to this subpart.

Subpart M—Floor and Wall Openings and Stairways

1926.500 Guardrails, handrails, and covers.
1910.21 Definitions.
1910.23 Guarding floor and wall openings and holes.
1926.501 Stairways.
1926.502 Definitions applicable to this subpart.

Subpart N—Cranes, Derricks, Hoists, Elevators, and Conveyors

1926.550 Cranes and derricks.
1926.551 Helicopters.
1926.552 Material hoists, personnel hoists, and elevators.
1926.553 Base-mounted drum hoists.
1926.554 Overhead hoists.
1926.555 Conveyors.
1926.556 Aerial lifts.

Subpart O—Motor Vehicles, Mechanized Equipment, and Marine Operations

1926.600 Equipment.
1910.169 Air receivers.
1910.176 Handling materials—general.
1926.601 Motor vehicles.
1926.602 Material handling equipment.
1926.603 Pile driving equipment.
1926.604 Site clearing.
1926.605 Marine operations and equipment.
1926.606 Definitions applicable to this subpart.

Subpart P—Excavations, Trenching, and Shoring

1926.650 General protection requirements.
1926.651 Specific excavation requirements.
1926.652 Specific trenching requirements.
1926.653 Definitions applicable to this subpart.

Subpart Q—Concrete, Concrete Forms, and Shoring

1926.700 General provisions.
1926.701 Forms and shoring.
1926.702 Definitions applicable to this subpart.

Subpart R—Steel Erection

1926.750 Flooring requirements.
1926.751 Structural steel assembly.
1926.752 Bolting, riveting, fitting-up, and plumbing-up.

Subpart S—Tunnels and Shafts, Caissons, Cofferdams, and Compressed Air

1926.800 Tunnels and shafts.
1926.801 Caissons.
1926.802 Cofferdams.
1926.803 Compressed air.
1926.804 Definitions applicable to this subpart.

Subpart T—Demolition

1926.850 Preparatory operations.
1926.851 Stairs, passageways, and ladders.
1926.852 Chutes.
1926.853 Removal of materials through floor holes.
1926.854 Removal of walls, masonry sections, and chimneys.
1926.855 Manual removal of floors.
1926.856 Removal of walls, floors, and material with equipment.
1926.857 Storage.
1926.858 Removal of steel construction.
1926.859 Mechanical demolition.
1926.860 Selective demolition by explosives.

Subpart U—Blasting and Use of Explosives

1926.900 General provisions.
1910.109 Explosives and blasting agents.
1926.901 Blaster qualifications.
1926.902 Surface transportation of explosives.
1926.903 Underground transportation of explosives.

1926.904 Storage of explosives and blasting agents.
1926.905 Loading of explosives or blasting agents.
1910.109 Explosives and blasting agents.
1926.906 Initiation of explosive charges—electric blasting.
1926.907 Use of safety fuse.
1926.908 Use of detonating cord.
1926.909 Firing the blast.
1926.910 Inspection after blasting.
1926.911 Misfires.
1926.912 Underwater blasting.
1926.913 Blasting in excavation work under compressed air.
1926.914 Definitions applicable to this subpart.

Subpart V—Power Transmission and Distribution

1926.950 General requirements.
1926.951 Tools and protective equipment.
1926.952 Mechanical equipment.
1926.953 Material handling.
1926.954 Grounding for protection of employees.
1926.955 Overhead lines.
1926.956 Underground lines.
1926.957 Construction in energized substations.
1926.958 External load helicopters.
1926.959 Lineman's body belts, safety straps, and lanyards.
1926.960 Definitions applicable to this subpart.

Subpart W—Rollover Protective Structures; Overhead Protection

1926.1000 Rollover protective structures (ROPS) for material handling equipment.
1926.1001 Minimum performance criteria for rollover protective structures for designated scrapers, loaders, dozers, graders, and crawler tractors.
1926.1002 Protective frame (ROPS) test procedures and performance requirements for wheel-type agricultural and industrial tractors used in construction.
1926.1003 Overhead protection for operators of agricultural and industrial tractors.

Subpart X—Effective Dates

1926.1050 Effective dates (general).
1926.1051 Effective dates (specific).

Supplement I—Text of other OSHA Standards incorporated by reference in those Part 1910 Standards identified as applicable to construction.

Safety Standards

1910.23(e)(11) *Wall opening screens.*
1910.106(b) *Tank storage.*
1910.106(c) *Piping, valves and fittings.*
1910.110(b) *Basic rules (LPG).*
1910.184(d) *Inspections (slings).*

Health Standards

1910.6 *Incorporation by reference.*
1910.100 *Standards organizations.*
1910.133 *Eye and face protection.*
1910.135 *Occupational head protection.*
1910.136 *Occupational foot protection.*
1910.141(d) (1), (2), (3) *Washing facilities.*
1910.141(e) *Change rooms*
1910.145(d)(4) *Caution signs.*

Supplement II—Table of "Threshold Limit Values of Airborne Contaminants for 1970" referenced in 29 CFR 1926.55(a) with administrative changes in footnotes and modifications to reflect subsequent OSHA rulemakings (Health Standards) issued pursuant to section 6(b) of the Act.

Supplement III—List of standards (other Federal agencies, ANSI, NEC, NFPA, etc.) incorporated by reference in Part 1926 and those part 1910 standards identified as applicable to construction work.

Private standards—setting organizations.
Federal agencies.
Other governmental agencies.

Supplement IV—List of addresses and telephone numbers (where appropriate) of Federal agencies, Private organizations, and other sources of standards references in OSHA standards applicable to construction work.

Addresses of standards organizations referenced.
Addresses of Federal agencies referenced.

Excerpts from Parts 1910 and 1926

For those contractors and managers new to the construction industry, several segments have been randomly selected from Part 1926 for illustration purposes. While reading over these excerpts, note the intermingling of Part 1910 with Part 1926.

Part 1910

SUBPART B—ADOPTION AND EXTENSION OF ESTABLISHED FEDERAL STANDARDS

1910.12—Construction Work (Extract)

(a) Standards

The standards prescribed in part 1926 of this chapter are adopted as occupational safety and health standards under section 6 of the Act and shall apply, according to the provisions thereof, to every employment and place of employment of every employee engaged in construction work. Each employer shall protect the employment and places of employment of each of his employees engaged in construction work by complying with the appropriate standards prescribed in this paragraph.

(b) Definition

For purposes of this section, "construction work" means work for construction, alteration, and/or repair, including painting and decorating. See discussion of these terms in § 1926.13 of this title.

(c) Construction Safety Act Distinguished

This section adopts as occupational safety and health standards under section 6 of the Act the standards which are prescribed in part 1926 of this chapter. Thus, the standards (substantive rules) published in subpart C and the following subparts of part 1926 of this chapter are applied. This section does not incorporate subparts A and B of part 1926 of this chapter. Subparts A and B have pertinence only to the application of section 107 of the Contract Work Hours and Safety Standards Act (the Construction Safety Act). For example, the interpretation of the term "subcontractor" in paragraph (c) of § 1926.13 of this chapter is significant in discerning the coverage of the Construction Safety Act and duties thereunder. However, the term "subcontractor" has no significance in the application of the Act, which was enacted under the Commerce Clause and which establishes duties for "employers" which are not dependent for their application upon any contracted relationship with the Federal Government or upon any form of Federal financial assistance.

(d) For the purposes of this part, to the extent that it may not already be included in paragraph (b) of this section, "construction work" includes the erection of new electric transmission and distribution lines and equipment, and the alteration, conversion, and improvement of the existing transmission and distribution lines and equipment.

Part 1926

SUBPART C—GENERAL SAFETY AND HEALTH PROVISIONS

1926.20—General Safety and Health Provisions

(a) Contractor Requirements

(1) Section 107 of the act requires that it shall be a condition of each contract which is entered into under legislation subject to Reorganization Plan Number 14 of 1950 (64 Stat. 1267), as defined in § 1926.12, and is for construction, alteration, and/or repair, including painting and decorating, that no contractor or subcontractor for any part of the contract work shall require any laborer or mechanic employed in the performance of the contract to work in surroundings or under working conditions which are unsanitary, hazardous, or dangerous to his health or safety.

(b) Accident Prevention Responsibilities

(1) It shall be the responsibility of the employer to initiate and maintain such programs as may be necessary to comply with this part.

(2) Such programs shall provide for frequent and regular inspections of the job sites, materials, and equipment to be made by competent persons designated by the employers.

(3) The use of any machinery, tool, material, or equipment which is not in compliance with any applicable requirement shall either be identified as unsafe by tagging or locking the controls to render them inoperable or shall be physically removed from its place of operation.

(4) The employer shall permit only those employees qualified by training or experience to operate equipment and machinery.

1926.21—Safety Training and Education

(a) General Requirements

The Secretary shall, pursuant to section 107(f) of the Act, establish and supervise programs for the education and training of employers and employees in the recognition, avoidance and prevention of unsafe conditions in employments covered by the act.

(b) Employer Responsibility

(1) The employer should avail himself of the safety and health training programs the Secretary provides.

(2) The employer shall instruct each employee in the recognition and avoidance of unsafe conditions and the regulations applicable to his work environment to control or eliminate any hazards or other exposure to illness or injury.

(3) Employees required to handle or use poisons, caustics, and other harmful substances shall be instructed regarding the safe handling and use, and be made aware of the potential hazards, personal hygiene, and personal protective measures required.

(4) In job site areas where harmful plants or animals are present, employees who may be exposed shall be instructed regarding the potential hazards, and how to avoid injury, and the first aid procedures to be used in the event of injury.

(5) Employees required to handle or use flammable liquids, gases, or toxic materials shall be instructed in the safe handling and use of these materials and made aware of the specific requirements contained in Subparts D, F, and other applicable subparts of this part.

(6)

(i) All employees required to enter into confined or enclosed spaces shall be instructed as to the nature of the hazards involved, the necessary precautions to be taken, and in the use of protective and emergency equipment required. The employer shall comply with any specific regulations that apply to work in dangerous or potentially dangerous areas.

(ii) For purposes of subdivision (i) of this subparagraph, "confined or enclosed space" means any space having a limited means of egress, which is subject to the accumulation of toxic or flammable contaminants or has an oxygen deficient atmosphere. Confined or enclosed spaces include, but are not limited to, storage tanks, process vessels, bins, boilers, ventilation or exhaust ducts, sewers, underground utility vaults, tunnels, pipelines, and open top spaces more than 4 feet in depth such as pits, tubs, vaults, and vessels.

1926.22—Recording and Reporting of Injuries [Reserved]

1926.23—First Aid and Medical Attention

First aid services and provisions for medical care shall be made available by the employer for every employee covered by these regulations. Regulations prescribing specific requirements for first aid, medical attention, and emergency facilities are contained in Subpart D of this part.

1926.24—Fire Protection and Prevention

The employer shall be responsible for the development and maintenance of an effective fire protection and prevention program at the job site throughout all phases of the construction, repair, alteration, or demolition work. The employer shall ensure the availability of the fire protection and suppression equipment required by Subpart F of this part.

1926.25—Housekeeping

(a) During the course of construction, alteration, or repairs, form and scrap lumber with protruding nails, and all other debris, shall be kept cleared from work areas, passageways, and stairs, in and around buildings or other structures.

(b) Combustible scrap and debris shall be removed at regular intervals during the course of construction. Safe means shall be provided to facilitate such removal.

(c) Containers shall be provided for the collection and separation of waste, trash, oily and used rags, and other refuse. Containers used for garbage and other oily, flammable, or hazardous wastes, such as caustics, acids, harmful dusts, etc., shall be equipped with covers. Garbage and other waste shall be disposed of at frequent and regular intervals.

1926.26—Illumination

Construction areas, aisles, stairs, ramps, runways, corridors, offices, shops, and storage areas where work is in progress shall be lighted with either natural or artificial illumination. The minimum illumination requirements for work areas are contained in Subpart D of this part.

1926.27—Sanitation

Health and sanitation requirements for drinking water are contained in Subpart D of this part.

1926.28—Personal Protective Equipment

(a) The employer is responsible for requiring the wearing of appropriate personal protective equipment in all operations where there is an exposure to hazardous conditions or where this part indicates the need for using such equipment to reduce the hazards to the employees.

(b) Regulations governing the use, selection, and maintenance of personal protective and lifesaving equipment are described under Subpart E of this part.

The following requirements from 29 CFR Part 1910 (General Industry) have been identified as applicable to construction (29 CFR 1926.28 —Personal Protective Equipment).

1910.132—General Requirements

(b) Employee-Owned Equipment

Where employees provide their own protective equipment, the employer shall be responsible to assume its adequacy, including proper maintenance, and sanitation of such equipment.

(c) Design

All personal protective equipment shall be of safe design and construction for the work to be performed.

SUBPART G—SIGNS, SIGNALS, AND BARRICADES

1926.200—Accident Prevention Signs and Tags

(a) General

Signs and symbols required by this subpart shall be visible at all times when work is being performed, and shall be removed or covered promptly when the hazards no longer exist.

(b) Danger Signs

(1) Danger signs (see Figure G-1) shall be used only where an immediate hazard exists.

Figure G-1

(2) Danger signs shall have red as the predominating color for the upper panel; black outline on the borders; and a white lower panel for additional sign wording.

(c) Caution Signs

(1) Caution signs (see Figure G-2) shall be used only to warn against potential hazards or to caution against unsafe practices.

Figure G-2

(2) Caution signs shall have yellow as the predominating color; black upper panel and borders; yellow lettering of "caution" on the black panel; and the lower yellow panel for additional sign wording. Black lettering shall be used for additional wording.

(d) Exit Signs

Exit signs, when required, shall be lettered in legible red letters, not less than 6 inches high, on a white field and the principal stroke of the letters shall be at least three-fourths inch in width.

(e) Safety Instruction Signs

Safety instruction signs, when used, shall be white with green upper panel with white letters to convey the principal message. Any additional wording on the sign shall be black letters on the white background.

(f) Directional Signs

Directional signs, other than automotive traffic signs specified in paragraph (g) of this section, shall be white with a black panel and a white directional symbol. Any additional wording on the sign shall be black letters on the white background.

(g) **Traffic Signs**

(1) Construction areas shall be posted with legible traffic signs at points of hazard.

(2) All traffic control signs or devices used for protection of construction workmen shall conform to American National Standards Institute D6.1-1971, Manual on Uniform Traffic Control Devices for Streets and Highways.

(h) **Accident Prevention Tags**

(1) Accident prevention tags shall be used as a temporary means of warning employees of an existing hazard, such as defective tools, equipment, etc. They shall not be used in place of, or as a substitute for, accident prevention signs.

(2) Specifications for accident prevention tags similar to those in Table G-1 shall apply.

(i) **Additional Rules**

American National Standards Institute (ANSI) Z35.1-1968, Specifications for Accident Prevention Signs, and Z35.2-1968, Specifications for Accident Prevention Tags, contain rules which are additional to the rules prescribed in this section. The employer shall comply with ANSI Z35.1-1968 and Z35.2-1968 with respect to rules not specifically prescribed in this subpart.

1926.201—Signaling

(a) **Flagmen**

(1) When operations are such that signs, signals, and barricades do not provide the necessary protection on or adjacent to a highway or street, flagmen or other appropriate traffic controls shall be provided.

(2) Signaling directions by flagmen shall conform to American National Standards Institute D6.1-1971, Manual on Uniform Traffic Control Devices for Streets and Highways.

TABLE G-1

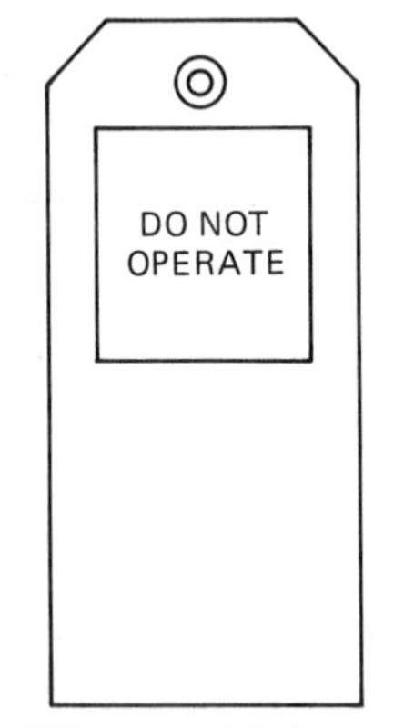

White tag — white letters on red square

White tag — white letters on red oval with a black square

Yellow tag — yellow letters on a black background

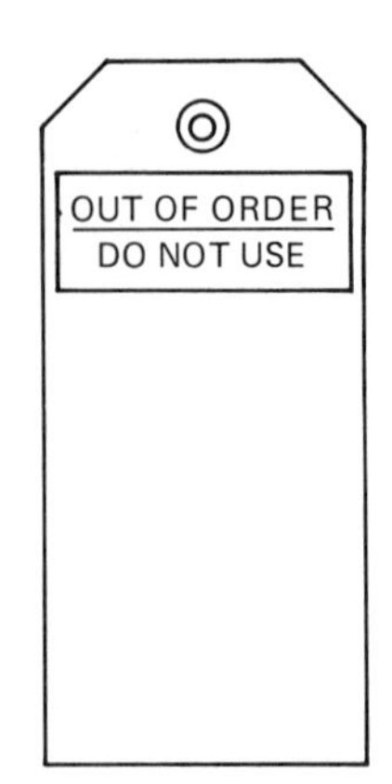

White tag — white letters on black background

Basic stock (background)	Safety colors (ink)	Copy specification (letters)
White	Red	Do not operate
White	Black and red	Danger
Yellow	Black	Caution
White	Black	Out of order — do not use

(3) Hand signaling by flagmen shall be by use of red flags at least 18 inches square or sign paddles, and in periods of darkness, red lights.

(4) Flagmen shall be provided with and shall wear a red or orange warning garment while flagging. Warning garments worn at night shall be of reflectorized material.

(b) Crane and Hoist Signals

Regulations for crane and hoist signaling will be found in applicable American National Standards Institute standards.

1926.202—Barricades

Barricades for protection of employees shall conform to portions of the American National Standards Institute D6.1-1971, Manual on Uniform Traffic Control Devices for Streets and Highways, relating to barricades.

1926.203—Definitions Applicable to This Subpart

(a) "Barricade" means an obstruction to deter the passage of persons or vehicles.

(b) "Signs" are the warnings of hazard, temporarily or permanently affixed or placed, at locations where hazards exist.

(c) "Signals" are moving signs, provided by workers, such as flagmen, or by devices, such as flashing lights, to warn of possible or existing hazards.

(d) "Tags" are temporary signs, usually attached to a piece of equipment or part of a structure, to warn of existing or immediate hazards.

SUBPART I—TOOLS—HAND AND POWER

1926.300—General Requirements

(a) Condition of Tools

All hand and power tools and similar equipment, whether furnished by the employer or the employee, shall be maintained in a safe condition.

(b) Guarding

(1) When power operated tools are designed to accommodate guards, they shall be equipped with such guards when in use.

(2) Belts, gears, shafts, pulleys, sprockets, spindles, drums, fly wheels, chains, or other reciprocating, rotating or moving parts of equipment shall be guarded if such parts are exposed to contact by employees or otherwise create a hazard. Guarding shall meet the requirements as set forth in American National Standards Institute, B15.1-1953 (R1958), Safety Code for Mechanical Power-Transmission Apparatus.

(c) Personal Protective Equipment

Employees using hand and power tools and exposed to the hazard of falling, flying, abrasive, and splashing objects, or exposed to harmful dusts, fumes, mists, vapors, or gases shall be provided with the particular personal protective equipment necessary to protect them from the hazard. All personal protective equipment shall meet the requirements and be maintained according to Subparts D and E of this part.

(d) Switches

(1) All hand-held powered platen sanders, grinders with wheels 2-inch diameter or less, routers, planers, laminate trimmers, nibblers, shears, scroll saws, and jigsaws with blade shanks one-fourth of an inch wide or less may be equipped with only a positive "on-off" control.

(2) All hand-held powered drills, tappers, fastener drivers, horizontal, vertical, and angle grinders with wheels greater than 2 inches in diameter, disc sanders, belt sanders, reciprocating saws, saber saws, and other similar operating powered tools shall be equipped with a momentary contact "on-off" control and may have a lock-on control provided that turnoff can

be accomplished by a single motion of the same finger or fingers that turn it on.

(3) All other hand-held powered tools, such as circular saws, chain saws, and percussion tools without positive accessory holding means, shall be equipped with a constant pressure switch that will shut off the power when the pressure is released.

(4) The requirements of this paragraph shall become effective on July 15, 1972.

(5) Exception: This paragraph does not apply to concrete vibrators, concrete breakers, powered tampers, jack hammers, rock drills, and similar hand operated power tools.

The following requirements from 29 CFR Part 1910 (General Industry) have been identified as applicable to construction (29 CFR 1926.300—General Requirements [Tools —Hand and Power]).

1926.301—Handtools

(a) Employers shall not issue or permit the use of unsafe handtools.

(b) Wrenches, including adjustable, pipe, end, and socket wrenches shall not be used when jaws are sprung to the point that slippage occurs.

(c) Impact tools, such as drift pins, wedges, and chisels, shall be kept free of mushroomed heads.

(d) The wooden handles of tools shall be kept free of splinters or cracks and shall be kept tight in the tool.

1926.304—Woodworking Tools

(a) Disconnect Switches

All fixed power driven woodworking tools shall be provided with a disconnect switch that can either be locked or tagged in the off position.

(b) Speeds

The operating speed shall be etched or otherwise permanently marked on all circular saws over 20 inches in diameter or operating at over 10,000 peripheral feet per minute. Any saw so marked shall not be operated at a speed other than that marked on the blade. When a marked saw is retensioned for a different speed, the marking shall be corrected to show the new speed.

(c) Self-Feed

Automatic feeding devices shall be installed on machines whenever the nature of the work will permit. Feeder attachments shall have the feed rolls or other moving parts covered or guarded so as to protect the operator from hazardous points.

(d) Guarding

All portable, power-driven circular saws shall be equipped with guards above and below the base plate or shoe. The upper guard shall cover the saw to the depth of the teeth, except for the minimum arc required to permit the base to be tilted for bevel cuts. The lower guard shall cover the saw to the depth of the teeth, except for the minimum arc required to allow proper retraction and contact with the work. When the tool is withdrawn from the work, the lower guard shall automatically and instantly return to the covering position.

(e) Personal Protective Equipment

All personal protective equipment provided for use shall conform to Subpart E of this part.

(f) Other Requirements

All woodworking tools and machinery shall meet other applicable requirements of American National Standards Institute, 01.1-1961, Safety Code for Woodworking Machinery.

1926.305—Jacks, Lever and Ratchet, Screw, and Hydraulic

(a) General Requirements

(1) The manufacturer's rated capacity shall be legibly marked on all jacks and shall not be exceeded.

(2) All jacks shall have a positive stop to prevent overtravel.

(b) Lift Slab Construction

(1) Hydraulic jacks used in lift slab construction shall have a safety device which will cause the jacks to support the load in any position in the event the jack malfunctions.

(2) If lift slabs are automatically controlled, a device shall be installed which will stop the operation when the ½-inch leveling tolerance is exceeded.

(c) Blocking

When it is necessary to provide a firm foundation, the base of the jack shall be blocked or cribbed. Where there is a possibility of slippage of the metal cap of the jack, a wood block shall be placed between the cap and the load.

1910.244—Other Portable Tools and Equipment

(a) Jacks [Material Deleted]

(2) Operation and maintenance

(iii) After the load has been raised, it shall be cribbed, blocked, or otherwise secured at once.

(iv) Hydraulic jacks exposed to freezing temperatures shall be supplied with an adequate antifreeze liquid.

(v) All jacks shall be properly lubricated at regular intervals. The lubricating instructions of the manufacturer should be followed, and only lubricants recommended by him should be used.

(vi) Each jack shall be thoroughly inspected at times which depend upon the service conditions. Inspections shall be not less frequent than the following:

(a) For constant or intermittent use at one locality, once every 6 months,

(b) For jacks sent out of shop for special work, when sent out and when returned,

(c) For a jack subjected to abnormal load or shock, immediately before and immediately thereafter.

(vii) Repair or replacement parts shall be examined for possible defects.

(viii) Jacks which are out of order shall be tagged accordingly, and shall not be used until repairs are made.

Concluding Comments

The information presented about OSHA is very sketchy. It was made that way deliberately to provide a very broad insight into the extent to which federal regulations protect the worker. An OSHA inspector may be called to any job site by any employee. Inspections will be made and action may be taken against any contractor who fails to comply. Without examining all the details, contractors do have certain rights regarding job-site inspections. If the reader is unfamiliar with the rights and responsibility he or she assumes, they would be well advised to research this subject further.

REFERENCES AND SOURCES

1. Educational sources
   a. College courses
      (1) Insurance

        (2) Safety management
        (3) OSHA
        (4) Personnel management
        (5) Labor relations
    b. Industrial seminars
    c. Insurance company seminars
2. Periodicals and books
    a. Safety
    b. Industrial safety
    c. Construction safety
    d. Building safety
    e. Building standards
    f. Southern building
3. Organizations dealing with the subjects:
    a. OSHA
    b. Building Officials and Code Administrators (BOCA), 17926 South Halsted Street, Homewood, IL 60430
    c. International Conference of Building Officials (ICBO), 5360 South Workman Mill Road, Whittier, CA 90601
    d. Southern Building Code Congress International (SBCCI), 900 Montclair Road, Birmingham, AL 35213
    e. Product, trade associations, and institutes (see Appendix B)

Appendix A

ARCHITECTURAL DRAWINGS

QUICK REFERENCES

■ DRAFTING LINES AND NOTES

Dimension Lines on Drawings

Definition. A dimension line is a fine dark line, the same width as an extension line. It is parallel to the section of the drawing being measured. It is continuous from extension line to extension line and has arrow tips at each end.

Rules

1. Dimension lines are usually spaced $\frac{3}{8}$ to $\frac{1}{2}$ inch from the section and the same distance from each other.
2. They are usually placed outside the section of the plan.

Types

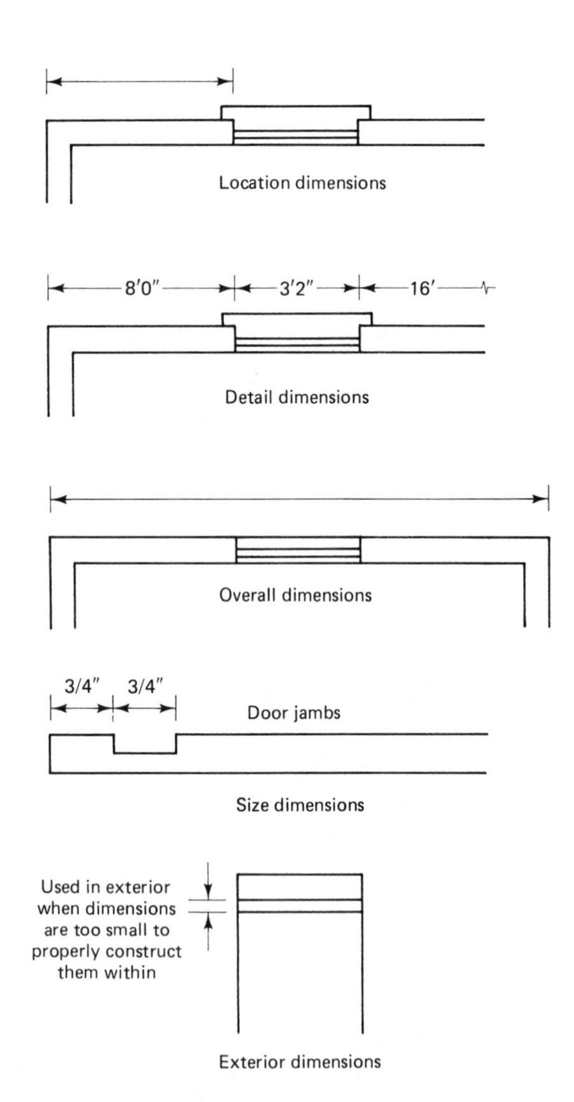

Location dimensions

Detail dimensions

Overall dimensions

Size dimensions

Exterior dimensions

Extension Lines on Drawings

Definition. An extension line is a fine dark line that extends at a right angle from an object, object line, or part to be dimensioned.

Rules

1. All extension lines begin $\frac{1}{16}$ inch from the object line and extend $\frac{1}{8}$ inch past the the outermost dimension line.
2. Use construction lines (very light, thin lines) extending at right angles from the object lines until the total number of dimension lines are determined.
3. Darken the portion of construction line that will be the finished extension line; erase the rest.

Examples

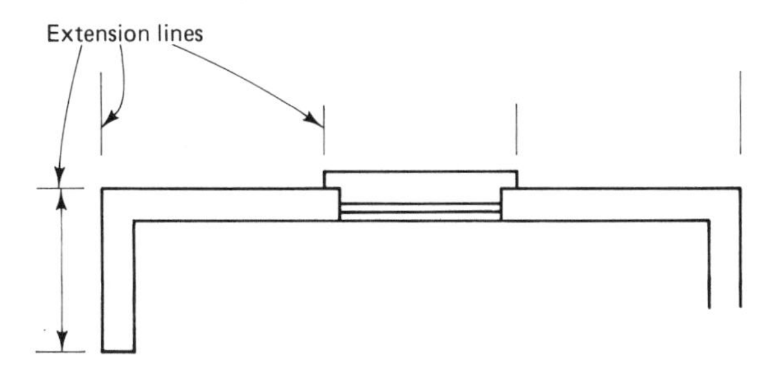

Object Lines

Definition. An object line on a drawing is a broadened, darkened construction line. Object lines are used for every part of the building that is an object (has mass): walls, sectional drawings, floor plan, windows, foundations.

Rules

1. Object lines do not extend through window or door openings on floor plans where an actual surface is not flush with the wall.
2. All equipment (heating, etc.) on a plan would be finalized with object lines.

Examples

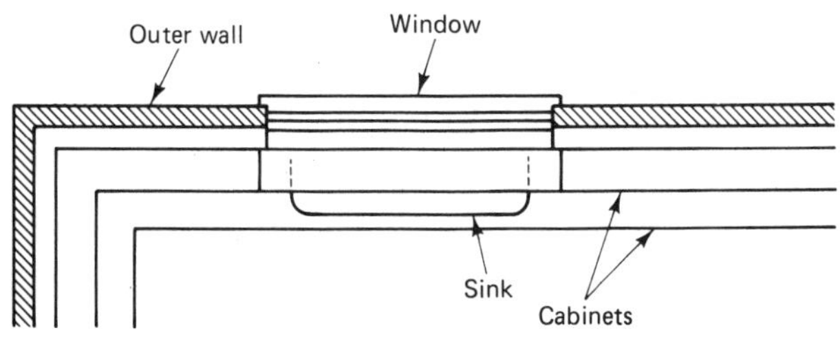

Object lines

Cutting-Plane Lines and Detail Callouts

Cutting-Plane Lines

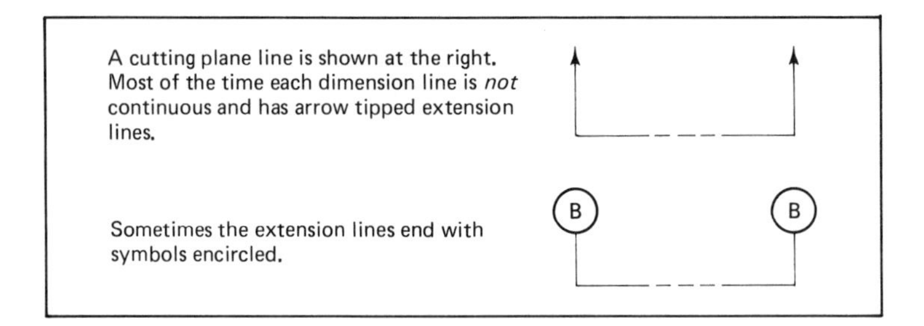

Detail Callouts

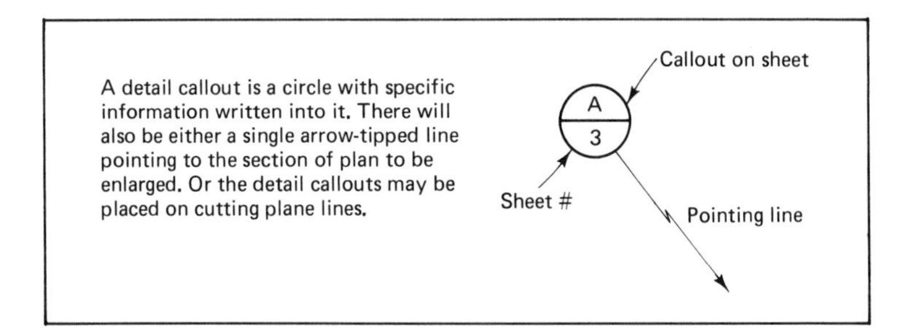

Notes

A list of notes is supplied on every drawing in the plan set where information is required but is too expensive or of insufficient usefulness to include on the plan if drawn in detail. For example, the window schedule is listed on the floor plan. Carpenters use the window sizes to prepare proper openings. A callout letter or note is indicated by each window on the plan. The contractor, estimator, and scheduler will use the same note for their own purposes. The contractor will need to know the types and sizes to assure proper installation. The estimator uses the information to cost the job. The scheduler will order the windows to arrive at the site on time.

■ ARCHITECTURAL PLAN LISTINGS FOR RESIDENTIAL CONSTRUCTION

Draftsmen number each sheet of the blueprint set in a standard manner. The list following is commonly used.

Sheet 1. Site Plan and Sheet Index
Sheet 2. First-Floor Plan
Sheet 3. Second-Floor Plan
Sheet 4. Basement or Foundation Plan
Sheet 5. Elevations
Sheet 6. Elevations
Sheet 7. Electrical
Sheet 8. Heating/AC
Sheet 9. Stair Detail
Sheet 10. Fireplace Detail
Sheet 11. Joist Framing Plan
Sheet 12. Roof Framing Plan
Sheet 13. Structural Details and Sections
Sheet 14. Structural Details and Sections
Sheet 15. Door and Window Schedule
Sheet 16. Room Finish Schedule

However, some offices may prefer a different numbering sequence. Simple homes, those with no special features, would require the least number of sheets, whereas an expensive contemporary home or a condominium may require many more sheets to illustrate special details.

ARCHITECTURAL PLAN LISTINGS FOR COMMERCIAL CONSTRUCTION

Since commercial projects are significantly different from residential types, plans contain many more details. For example, most commercial buildings use fireproof materials, heavier materials, and steel superstructors or steel reinforcements. Therefore, a greater variety of workers will probably be employed. These people will need detailed plans for their part in the construction of the project.

As with residential plans, the sheets may be numbered differently by different architects. A sample sheet numbering schedule follows:

Sheet 1. Sit Plan and Sheet Index
Sheet 2. Key Plan
Sheet 3. Roof Plan
Sheet 4. Foundation Plan
Sheet 5. Floor Plan
Sheet 6. Room Finish Schedule
Sheet 7. Door Schedule
Sheet 8. Frame Schedule
Sheet 9. Exterior Elevations
Sheet 10. Interior Elevations
Sheet 11. Building Sections
Sheet 12. Details and Sections
Sheet 13. Casework Details
Sheet 14. Special Equipment Plans
Sheet E-1. Electrical
Sheet M-1. Mechanical Plan (Plumbing and Heating)
Sheet S-1. Roof Framing Plan

ARCHITECTURAL DRAWING SYMBOLS (TRADE OR MATERIAL SYMBOLS)

Illustrated below and on p. 181 are symbols used for:

1. General materials

MATERIAL	PLAN	ELEVATION	SECTION
Brick	Common Face Firebrick	Same as Common Brick Same as Above	Same as Plan View
Stone	Cut Stone Rubble Cast Stone (Concrete)	Cut Stone Rubble	Same as Plan View
Concrete	Concrete or Concrete Block	Concrete Concrete Block	Same as Plan View
Structural Steel	or or	None	or
Interior Partitions	Studs, Lath and Plaster Solid Plaster Wall		Same as Plan View
Glass		or	Small Scale Large Scale
Insulation	Loose Fill, or Batts Board and Quilt Solid and Cork	None	Same as Plan View

MATERIAL	PLAN	ELEVATION	SECTION
Wood	Floor Areas Left Blank Indicates Kind of Wood Used	Siding Panel	End of Board (Except Trim) Trim
Sheet Metal Flashing	Indicate by Note		Heavy Line Shaped To Conform
Earth	None	None	
Rock	None	None	
Sand	None	None	
Gravel or Cinders	None	None	
Floor and Wall Tile			
Soundproof Wall		None	None
Plastered Arch		Design Varies	Same as Elevation View
Glass Block in Brick Wall			Same as Elevation View
Brick Veneer	or On Frame On Concrete Block	Same as Brick	Same as Plan View
Cut Stone Veneer	or On Frame On Brick On Concrete Block	Same as Cut Stone	Same as Plan View
Rubble Stone Veneer	or On Frame On Brick On Concrete Block	Same as Rubble	Same as Plan View

2. Plumbing and heating/air conditioning (below and on p. 183)

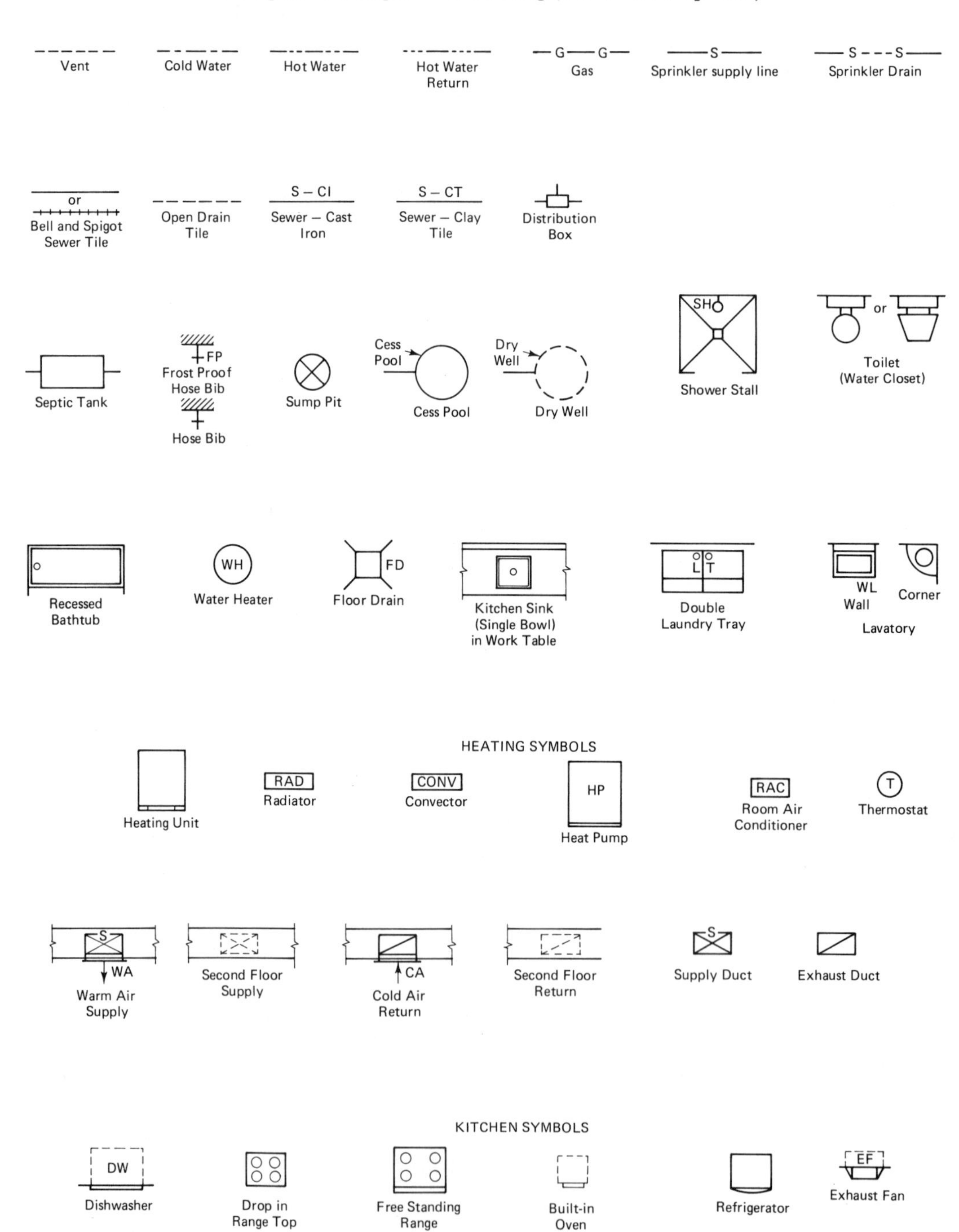

SUPPLY LINES

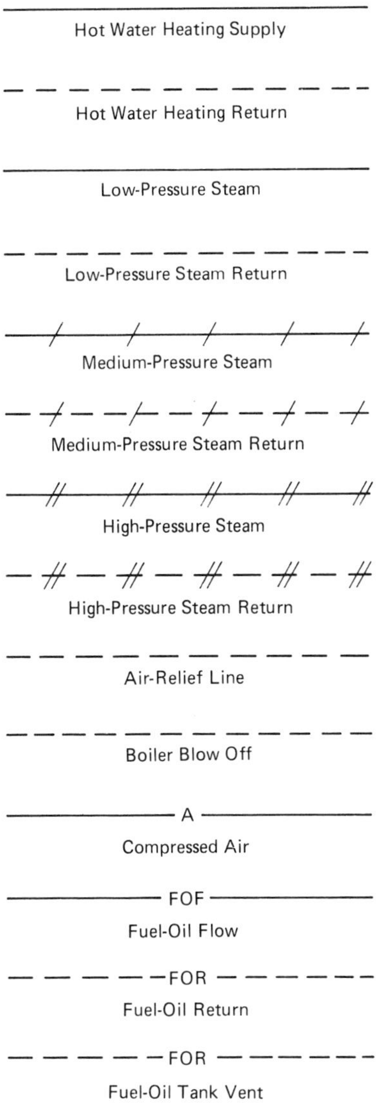

Hot Water Heating Supply

Hot Water Heating Return

Low-Pressure Steam

Low-Pressure Steam Return

Medium-Pressure Steam

Medium-Pressure Steam Return

High-Pressure Steam

High-Pressure Steam Return

Air-Relief Line

Boiler Blow Off

A

Compressed Air

FOF

Fuel-Oil Flow

FOR

Fuel-Oil Return

FOR

Fuel-Oil Tank Vent

HEATING/AC DUCTS AND REGISTERS

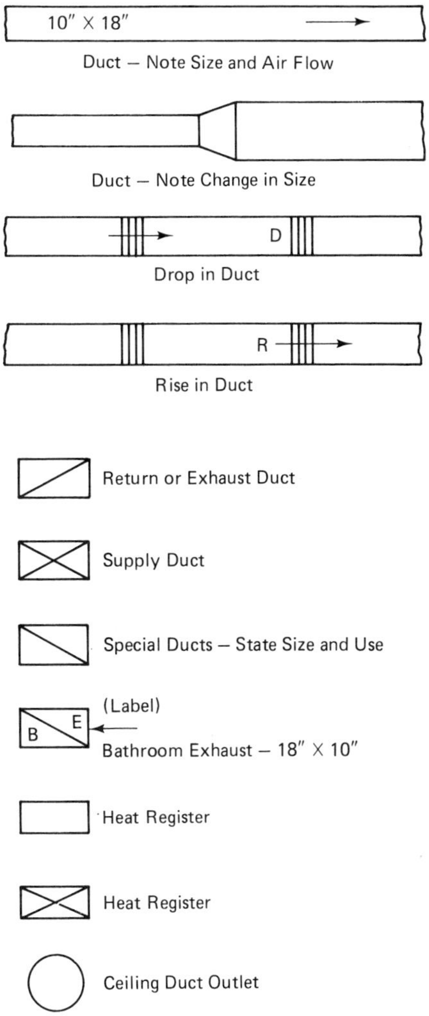

10″ X 18″

Duct – Note Size and Air Flow

Duct – Note Change in Size

D

Drop in Duct

R

Rise in Duct

Return or Exhaust Duct

Supply Duct

Special Ducts – State Size and Use

(Label)

B E

Bathroom Exhaust – 18″ X 10″

Heat Register

Heat Register

Ceiling Duct Outlet

3. Electrical systems (below and on p. 185)

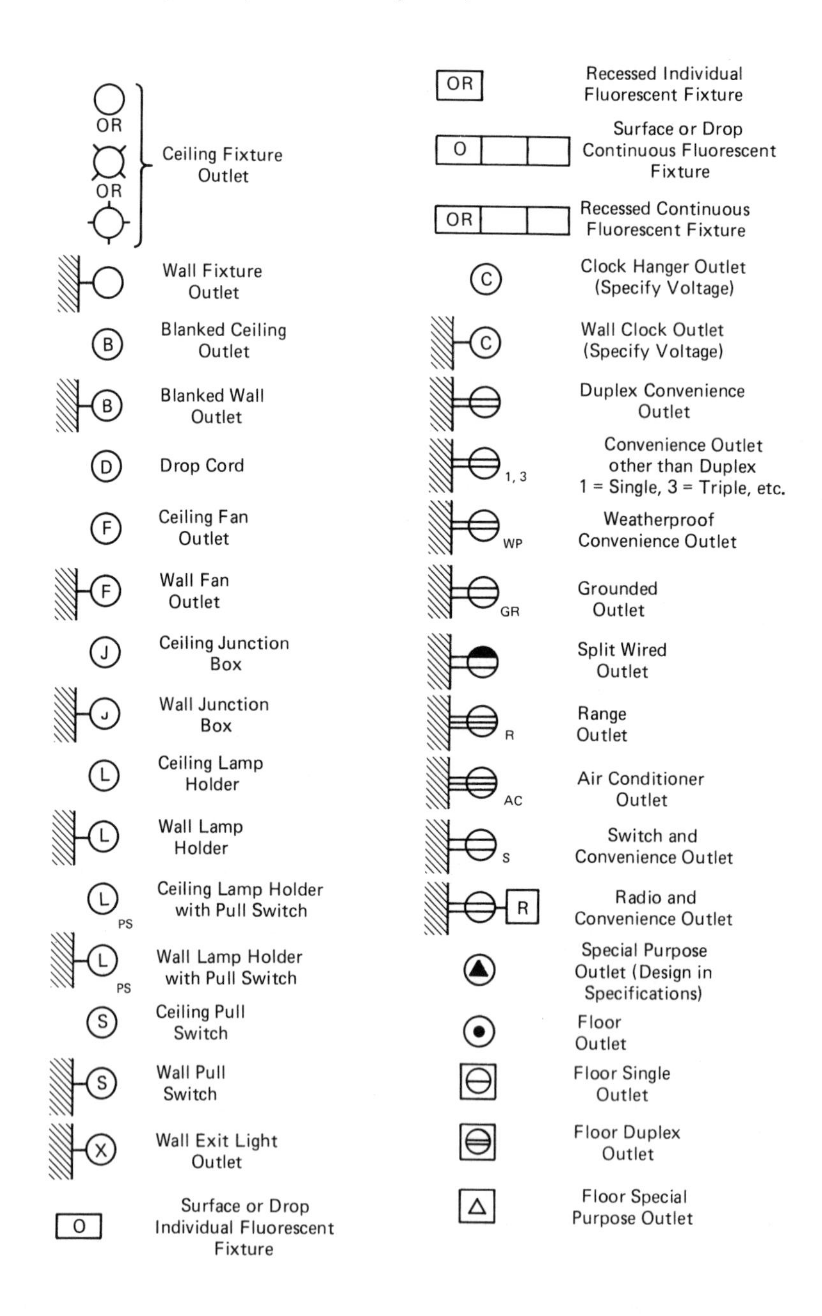

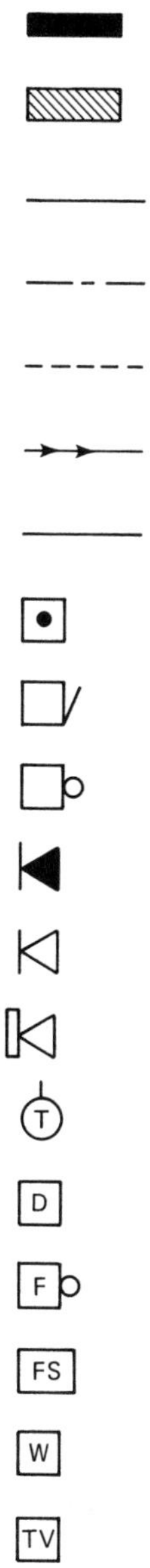

- Lighting Panel
- Power Panel
- Branch Circuit; Concealed in Ceiling or Wall
- Branch Circuit; Concealed in Floor
- Branch Circuit; Exposed
- Home Run to Panel Board; Indicate Number of Circuits by Number of Arrows
- Feeders
- Push Button
- Buzzer
- Bell
- Outside Telephone
- Interconnecting Telephone
- Telephone Switch Board
- T — Bell Ringing Transformer
- D — Electric Door Opener
- F — Fire Alarm Bell
- FS — Automatic Fire Alarm Device
- W — Watchman's Station
- TV — TV Outlet

a, b, c

a, b, c

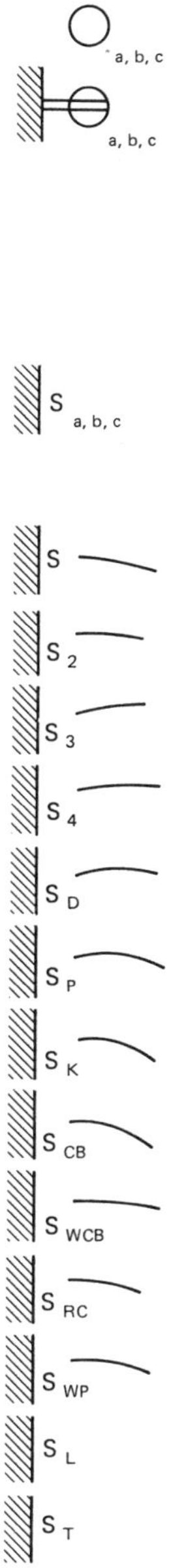

Any Standard Symbol Given with the Addition of a Lower Case Subscript may be Used to Designate Some Special Variation of Standard Equipment of Particular Interest in a Set of Architectural Plans.

$S_{a, b, c}$

When Used They Must Be Listed in the Key of Symbols on Each Drawing and if Necessary Further Described in the Specifications.

- S — Single Pole Switch
- S_2 — Double Pole Switch
- S_3 — Three Way Switch
- S_4 — Four Way Switch
- S_D — Automatic Door Switch
- S_P — Switch and Pilot Lamp
- S_K — Key Operated Switch
- S_{CB} — Circuit Breaker
- S_{WCB} — Weatherproof Circuit Breaker
- S_{RC} — Remote Control Switch
- S_{WP} — Weatherproof Switch
- S_L — Low Voltage Switch
- S_T — Time Switch

4. Windows and doors

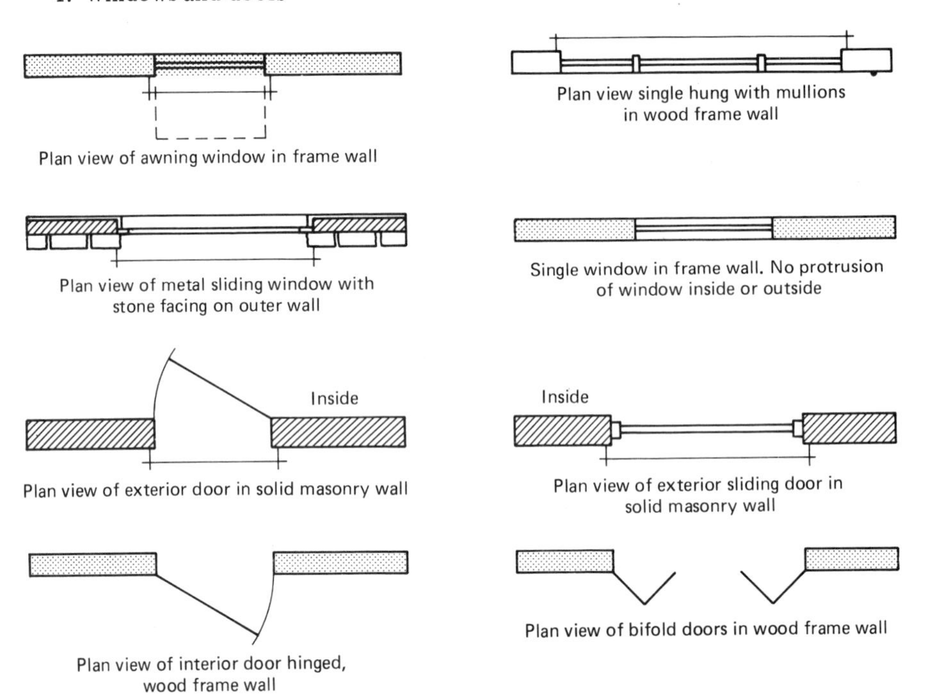

Plan view of awning window in frame wall

Plan view single hung with mullions in wood frame wall

Plan view of metal sliding window with stone facing on outer wall

Single window in frame wall. No protrusion of window inside or outside

Plan view of exterior door in solid masonry wall

Plan view of exterior sliding door in solid masonry wall

Plan view of interior door hinged, wood frame wall

Plan view of bifold doors in wood frame wall

ABBREVIATIONS AND TERMS

Space on drawings and in the list of specifications is always at a premium. Qualified contractors and journeymen do not have time to read, nor do they care to read, lengthy terms. Therefore, abbreviations, or more commonly referred symbols, are used in place of fully written out terms.

Each part of the construction industry has a set of abbreviations and terms unique to its particular scope. For example, Exhibits A-1 and A-2 show three different parts. Exhibit A-1 shows some of the abbreviations used on architectural plans in general. Exhibit A-2 shows lumber abbreviations frequently found in specification listings. Each material type (e.g., masonry, glass, steel, and wood) and each trade has a set of abbreviations that is uniquely theirs.

A good reference book for abbreviations and terms is *Architectural Graphic Standards* by Ramsey and Sleeper.[1]

Specialized trade symbols can also be obtained from numerous trade organizations or associations or institutions closely associated with particular trades.

[1]C. G. Ramsey and H. R. Sleeper, *Architectural Graphic Standards*, 7th ed. (New York: Wiley, 1980).

Exhibit A-1 Architectural graphic abbreviations and symbols.

A: Asphalt
A̲: Surface area
A B: Anchor bolt
A C or a-c: Alternating current
ACST or AC: Acoustic or acoustical
A D: Area drain
AGGR: Aggregate
AL: Aluminum
ALM: Alarm
ALT: Altitude
ARCH: Architect
ASB: Asbestos
ASPH: Asphalt
A T: Asphalt tile
AVE: Avenue
Bar ⌀: Round bar
BAR □: Square bar
BATH: Bathroom
BD. FT: Board foot
BET: Between
BLK: Block
BLKG: Blocking
BLR: Boiler
B M: Beam; bench mark
B R: Bedroom; bottom register
BSMT: Basement
B T: Bathtub
BT: Bolts
B T U or Btu: British thermal units
B U R: Built-up roof
C: Channel; channel, American standard; closet, courses
CAB: Cabinet
C B: Catch basin
C CONC: Cast concrete
CEM: Cement
CEM A: Cement asbestos
CEM AB: Cement asbestos board
CER: Ceramic
CIN BL: Cinder block
CIR BKR: Circuit breaker
CL: Center line; closet
CLG: Ceiling
CLO: Closet
CLR: Clear
cm: Centimeter
CM: Center matched
CO: Cleanout
CO D: Cleanout door
COL: Column
COM or Com: Common
CONC: Concrete
CONC B: Concrete block
COND: Conductor
CONST: Construction
CONTR: Contractor
C R: Ceiling register; center register
CRS: Cold-rolled steel
C S: Cast stone
C to C or c to c: Center to center
CTR: Center
CU or cu: Cubic
CU YD or cu yd: Cubic yard
CW: Cold water
D: Dawn; drain, dryer
D A: Double acting
DET: Detail
DF: Drinking fountain
D H: Double hung
D H W: Domestic hot water
DIA: Diameter
DIM or dim: Dimension
DKG: Decking
DMPR: Damper
DN: Down
DP: Dampproofing
D R: Dining room
DR: Drain
DS: Downspout
DUP: Duplicate
DVTL: Dovetail
D W: Dry well
DW: Dishwasher

DX: Duplex
E: East
EL or el: Elevation
ELEC: Electric
ENCL: Enclose
ENT: Entrance
EST: Estimate
EXC: Excavate
EXP BT: Expansion bolt
EXP JT: Expansion joint
EXT: Exterior
F BRK: Firebrick
FD: Floor drain
F EXT: Fire extinguisher
F HC: Fire hose cabinet
FIN: Finish
FIN FL: Finish floor
FIX: Fixture
FL: Flashing; floor; flush
FLG or Flg: Flooring
FLUOR: Fluorescent
FP: Fireplace
FT or ft: Feet or foot
FTG: Footing
G: Granite; grille; gypsum
GA: Gage
GALV: Galvanized
GD: Guard
GEN CONT: General contractor
G I: Galvanized iron
G L: Grade line
GL: Glass; glaze
GL BL: Glass block
GR: Game room; grade
GRTG: Grating
G S: Gravel stop
G T: Grease trap
H: Height
HD: Head
HDW: Hardware
HDWD or Hdwd: Hardwood
H F or H B: Hose faucet or bib
HGT: Height
HOR: Horizontal
H PT: High point
HSE: House
HT: Height
HTR: Heater
HW: Hot water
H WH: Hot water heater
I: Iron
I D: Inside diameter
IN: Inch
INS: Insulate or insulation
INT: Interior
J CL: Janitor's closet
JT: Joint
K: Kitchen
KM: Kilometer
KWH or kwhr: Kilowatt hour
L: Lead; leader; length
LAD: Ladder
LAU: Laundry
LAV: Lavatory
LB or lb: Pound
LB/FT^2: Pounds per square foot
LBR or lbr: Lumber
L C: Laundry chute
L CL: Linen closet
L D: Leader drain
LDG: Landing
LG: Length
lgth: Length
L H: Left hand
LIB: Library
LKR: Locker
LN: Lane
L O: Louver opening
L P: Low point
L R: Living room
LS: Limestone
LT: Light

LTH: Lath
LV: Louver
LV D: Louvered door
LV O: Louver opening
LWB: Lightweight block
LWC: Lightweight concrete
m: Meter (measure)
M: Thousand
MATL: Material
MBM: 1000 board feet
M C: Junior channel; medicine cabinet
MECH: Mechanical
MET: Metal
MEZZ: Mezzanine
MFG: Manufacturing
MIN: Minimum
MISC: Miscellaneous
MLDG: Moulding
mm: Millimeter
M O: Masonry opening
MR: Marble
N: North; number
NO: number
NOM: Nominal
O C: On center
O D: Outside diameter
OFF: Office
OPNG: Opening
OPP: Opposite
OUT: Outlet
OV HD: Overhead
P: Plate; pull chain
PAN: Pantry
PASS: Passage
P C: Pull chain
PL: Plaster; plate
PL GL: Plate glass
PLAS: Plaster
PLMB: Plumbing
PLSTC: Plastic
PNL: Panel
PORT: Portable
PSF: Pounds per square foot
PTD: Painted
PTN: Partition
P V C: Polyvinyl chloride
Q T: Quarry tile
Q T B: Quarry tile base
Q T F: Quarry tile floor
R: Range; riser
Rd: Road
R D: Roof drain
RD: Round
REF: Refrigerator
REFL: Reflective
REG: Register; regulator
RET: Return
REV: Revision
RF: Roof
RFG or Rfg: Roofing
RGH: Rough
R H: Right hand
RIV: Rivet
R/L: Random length
rnd: Round
R P M or rpm: Revolutions per minute
S: I beam; sewer; sink; south; surface area; switch
SC: Scale
SCH: Schedule
SDG or Sdg: Siding
SEC: Section (land)
SECT: Section (drafting)
SERV: Service
SFTWD: Softwood
SH: Shelves; shower
SHTHG: Sheathing
SK: Sink; switch, key operated
SL: Slate
S P: Soil pipe; sump pit
SPEC: Specifications
SPKR: Speaker

SPR: Sprinkler
SQ: Square
sq ft: Square foot
sq in: Square inch
SS: Service sink
S ST: Stainless steel
ST: Plumbing stack; stairs; street
STD: Standard
STG: Storage
STN: Stone
STR: Structural
STWY: Stairway
SUB: Substitute
SUP: Supply
SUR: Surface
SUSP CEIL: Suspended ceiling
SW: Switch
T: Tee; thick or thickness, toilet, transom; tread
T & G: Tongue & groove
TAN: Tangent
T C: Terra cotta
TEL: Telephone
TEMP: Temperature
TER: Terrazzo
THERMO: Thermostat
THK: Thick or thickness
TR: Top register
TR: Transom; tread
T S: Tubing, structural
U: Up
UNFIN: Unfinished
UR: Urinal
V: Vent; ventilator; volume
VAP PRF: Vapor proof
VARN: Varnish
V D: Vent duct
VENT: Ventilate
VERT: Vertical
VEST: Vestibule
VOL: Volume
V P: Vest pipe
V S: Vent stack
V T: Vinyl tile
V T B: Vinyl tile base
V T F: Vinyl tile floor
W: Wall; watt; west; wide flange; width
W C: Water closet
W CAB: Wall cabinet
WD: Wood; wood door
WDW: Window
WF: Wood frame
W GL: Wire glass
W H: Water heater; hot water heater
WH: Weephole
WI: Wrought iron
WM: Washing machine
W P: Water proofing
WP: Weatherproof
WR: Washroom
W S: Weather stripping
W T: Water table
WT: Watertight
WT or wt: Weight
wth: Width
W V: Wall vent
X H: Extra heavy
X HVY: Extra heavy
YD or yd: Yard
Z: Zinc

SYMBOLS

∠: Angle
L: Angle iron; equal leg angle
∠: Unequal leg angle
@: At
x: By
℄: Centerline
′: Feet or foot
″: Inch
#: Number; pound

Exhibit A-2 Lumber abbreviations and symbols.

AD: Air-dried
ADF: After deduction freight
ALS: American Lumber Standards
AVG: Average
AW&L: All widths and lengths
BD: Board
BD. FT: Board feet
BDL: Bundle
BEV: Bevel
BH: Boxed heart
B/L, BL: Bill of lading
BM: Board measure
B&S: Beams and stringers
BSND: Bright sapwood, no defect
BTR: Better
CB: Center beaded
C/L: Carload
CLG: Ceiling
CLR: Clear
CM: Center matched
CS: Caulking seam
CSG: Casing
CV: Center V
DET: Double end trimmed
DF: Douglas fir
DF-L: Douglas fir-larch
DIM: Dimension
DKG: Decking
D&M: Dressed and matched
D/S, DS: Drop siding
E: Edge or modulus of elasticity
EB1S: Edge bead one side
EB2S: Edge bead two sides
E&CB2S: Edge & center bead two sides
EV1S: Edge vee one side
EV2S: Edge vee two sides
E&CV1S: Edge and center vee, one side
E&CV2S: Edge and center vee, two sides
EE: Eased edged
EG: Edge (vertical) grain
EM: End matched
ES: Engelmann spruce
f: Allowable fiber stress in bending (also Fb)
FG: Flat or slash grain
FLG: Flooring
FOHC: Free of heart center
Ft: Foot
FT. BM: Feet board measure (also FBM)
FT. SM: Feet surface measure
H.B.: Hollow back
HEM: Hemlock
H&M: Hit and miss
H or M: Hit or miss
IC: Incense cedar
IN: Inch or inches
IND: Industrial
IWP: Idaho white pine
J&P: Joists and planks
JTD: Jointed
KD: Kiln-dried
L: Larch
LBR: Lumber
LF: Light framing
LGR: Longer
LGTH: Length
LIN: Lineal
LNG: Lining
LP: Lodgepole pine
M: Thousand
MBF: Thousand board feet
M. BM: Thousand (ft) board measure
MC: Moisture content
MG: Mixed grain
MLDG: Moulding
MOE: Modulus of elasticity or *E*

MOR: Modulus of rupture
MSR: Machine stress rated
NBM: Net board measure
N1E: Nose one edge
PAD: Partly air dried
PART: Partition
PAT: Pattern
PET: Precision end trimmed
PP: Ponderosa pine
P&T: Posts and timbers
RC: Red cedar
RDM: Random
REG: Regular
RGH: Rough
R/L, RL: Random lengths
R/S: Resawn
R/W, or RW: Random widths
R/W, R/L: Random width, Random length
SB1S: Single bead one side
SDG: Siding
S&E: Side and edge
S1E: Surfaced one edge
S2E: Surfaced two edges
S1S: Surfaced one side
S2S: Surfaced two sides
S4S: Surfaced four sides
S1S&CM: Surfaced one side and center matched
S2S&CM: Surfaced two sides and center matched
S4S&CS: Surfaced four sides and caulking seam
S1S1E: Surfaced one side, one edge
S1S2E: Surfaced one side, two edges
S2S1E: Surfaced two sides, one edge
SEL: Select
SG: Slash or flat grain
S/L or SL: Shiplap
SM: Surface measure
SP: Sugar pine
SQ: Square
STD.M.: Standard matched
STK: Stock
STPG: Stepping
STR: Structural
TBR: Timber
T&G: Tongued and grooved
VG: Vertical (edge) grain
WDR: Wider
WF: White fir
WRC: Western red cedar
WT: Weight
WTH: Width
WWPA: Western Wood Products Association

SYMBOLS

″: Inch or inches
′: Foot or feet
x: By, as 4 x 4
4/4, 5/4, 6/4, etc.: nominal thickness expressed in fractions

Appendix B

OTHER SOURCES OF INFORMATION

QUICK REFERENCES

Almost every one of the sources identified in this appendix can be found in a well-stocked public library.

■ INSTITUTIONS AND ASSOCIATIONS

One of the most comprehensive listings of trade and construction industry institutions and associations is the *Encyclopedia of American Associations*. There are about 9000 listings, including professional societies, nonprofit organizations, and chambers of commerce. The directory includes accounting, architectural, business, commercial organizations, governmental, legal, scientific, engineering, technical, educational, trade, and other categories. It can be found in almost every library and is also available from the publisher at the following address:

Gale Research Co.
34th Floor, Book Tower
Detroit, MI 48226

UNIVERSITIES

Many universities are instrumental in developing modern and new techniques and building materials. They have extensive laboratories and departments. Much of this information is available free or at nominal cost. The reports may be extensive bulletins or only one page in length.

A description of the activities of the universities that have research departments can be found in the *Directory of University Research Bureaus and Institutes*. This directory is also available in many libraries and from Gale Research Co.

BUILDING CODE ORGANIZATIONS

1. *National Building Code:* American Insurance Association, 85 John Street, New York, NY 10038 (this code is used in the eastern states).
2. *Basic Building Code:* Building Officials and Code Administrators International, 17926 South Halsted Street, Homewood, IL 60430 (this code is used in the midwestern and northeastern states).
3. *Uniform Building Code:* International Conference of Building Officials, 5360 South Workman Mill Road, Whittier, CA 90601 (this code is used in the western states).
4. *Southern Standard Building Code:* Southern Building Code Congress, 900 Montclair Road, Birmingham, AL 35213

TRADE AND PRODUCT SOURCES

The *Thomas Register of American Manufacturers* is a most complete listing of manufacturers available. Almost every library has a copy of this multivolume register. There is a section containing an alphabetical listing of each manufacturer and its products. There is also a section organized by product. Each volume contains an index and trade name directory that makes locating a manufacturer or product simple to find. The register can be ordered from the publisher at the following address:

Thomas Publishing Co.
461 Eighth Avenue
New York, NY 10001

Sweet's Catalog Service is a compilation of manufacturers' catalogs, classified by type of product. The catalog is contained in more than one volume. Each type of product is cataloged under Architectural, Industrial Construction, or Light Construction. The catalogs are well indexed and contain a numbering system for ease of location. *Sweet's Catalog* can be found in almost every building supply house and can be ordered from the publisher, whose address is:

F.W. Dodge Corporation
330 West 42nd Street
New York, NY 10036

GOVERNMENT SOURCES

1. Superintendent of Documents, U.S. Government Printing Office, Washington, DC 20402 (send for listings on specific subjects).
2. Business Service Center, U.S. Department of Commerce, Washington, DC 20230 (use this source when a governmental agency is needed to help supply information on a problem).
3. U.S. and state departments of:
   a. Agriculture (big on housing)
   b. Housing or housing and urban development (HUD)
   c. Education
   d. Energy (involved in solar)

   Write to the Department in Washington, DC, for federal information or to the state capital for state information.
4. Internal Revenue Service:
   a. Local service centers (see the phone book)
   b. Regional service centers (see the phone book)

OTHER SOURCES

1. *How and Where to Look It Up* tells how to use libraries, encyclopedias, almanacs, guides to current literature, and other sources of information. It is available in most libraries and from the publisher:

   McGraw-Hill Book Company
   1221 Avenue of the Americas
   New York, NY 10020

2. *How and Where to Find Facts* is an encyclopedia that gives various sources of information on a large variety of subjects. This book is also available in many libraries and from the publisher:

   Arco Publishing, Inc.
   480 Lexington Avenue
   New York, NY 10017

3. *Sources of Business Information* by Edwin J. Comon lists business libraries, government sources, statistical and financial sources and many specialized sources on building and materials. It can be ordered from the publisher at the following address:

   Prentice-Hall, Inc.
   Englewood Cliffs, NJ 07632

INDEX